持続可能性と戦略

企業と社会フォーラム 編

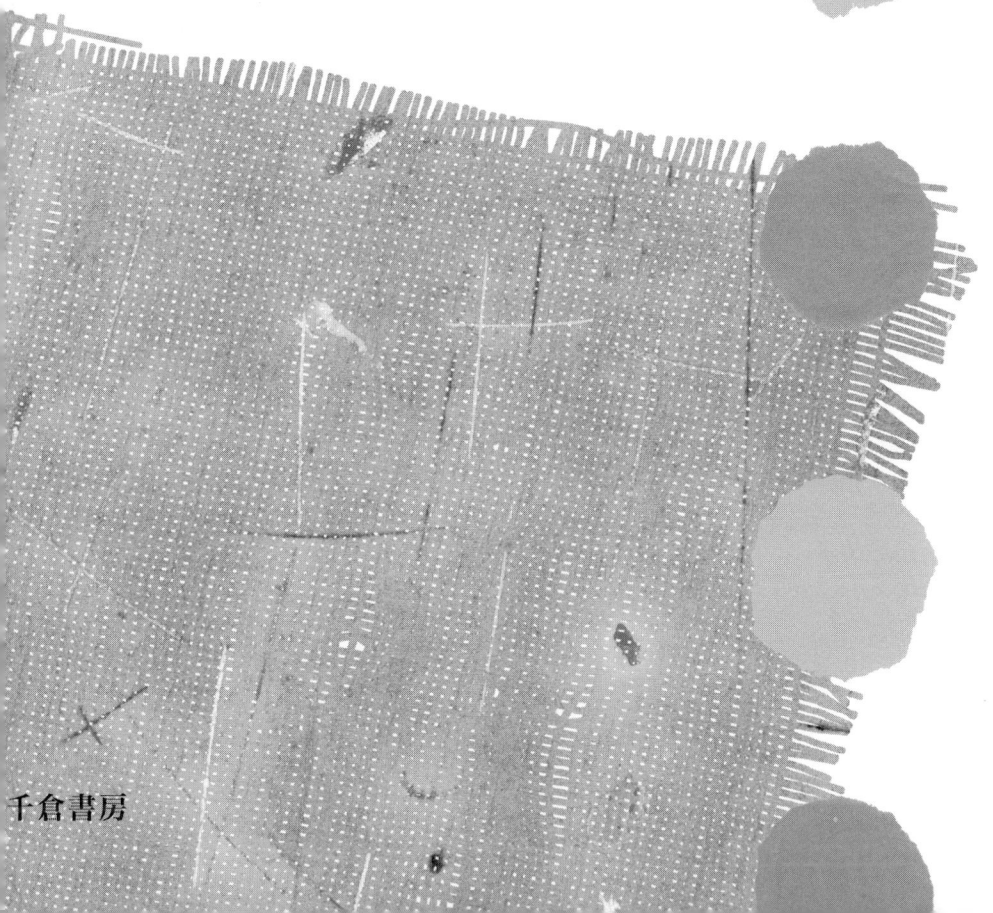

千倉書房

は じ め に

　本書は「持続可能性と戦略」をテーマに，理論的・実践的な視点から考えていくことを目的に編纂されている。
　持続可能性（sustainability）という概念は，環境問題のみならず，社会問題を含め，経済，環境，社会をトータルに捉え発展していくことが重要であるとの理解に基づいている。企業経営の現場でも近年，CSR をいかに経営プロセスに組み込んでいくか，さらに持続可能な発展（sustainable development）に貢献する新しい事業戦略が求められるようになっている。経営戦略論の領域でもこういった動きを捉え，分析していくことが求められはじめている。例えば，Strategic Management Society が 2013 年に開催した年次大会では，Strategy and Sustainability を統一テーマとし，議論している。
　これまで経営戦略論において持続可能性というと，「持続可能な競争優位」（sustainable competitive advantage）という理解のされ方がなされていた。そこでは企業の戦略行動がもたらす環境・社会への影響に関する問題を扱うことはなかった。もっぱら経済の問題として取り扱われ，企業の目的は株主資本価値の最大化であった。
　しかし現在，実務家も経営戦略論の研究者も，市場環境が大きく変わり，企業の持続可能性は経済合理性に基づいて経営資源を最適配分するだけでは実現できないということを理解しはじめている。株主価値の最大化も，経済・環境・社会をトータルに捉えることがなければ達成されないし，環境的・社会的課題の解決に貢献するイノベーションが新しい価値を生む。そういった思考は，これまでとは異なるビジネスモデルや，組織文化，経営手法から生まれてくる。研究面では，新たな理論とそれを検証する研究手法の開発が求められている。

　「企業と社会フォーラム」は，こういった課題を議論するため，「持続可能性と戦略」を統一テーマとし，第 4 回年次大会を 2014 年 9 月 18 日（木）～19 日（金）の 2 日間にわたり，早稲田大学にて開催した。そこではアジア・太平洋を中心に 8 カ国／地域（アメリカ，イギリス，オーストラリア，ニュージーランド，

台湾，中国，ドイツ，日本）の学界，産業界，労働界，NPO/NGO，学生など約100名が参加し，議論および交流を行った。

　本大会では岡田正大委員長（慶応義塾大学教授）を中心にProgram Committeeを設置し，Nick Barter（Senior Lecturer, Griffith University, Australia），Philippe Debroux（創価大学教授），谷本寛治（早稲田大学教授）が，大会報告プロポーザルの審査，企画セッションやDoctoral Workshopの司会を担当した。

　本書は，この第4回年次大会での議論を踏まえ，その後の研究成果や企業事例報告をとりまとめたものである。第Ⅰ部では，本テーマに関する「招待論文」5本を収めている。その内，キーノートスピーカーであったBhattacharya教授の論文については，JFBS事務局で翻訳を行い掲載している。第Ⅱ部「企業ケース」では，積水ハウス，NEC，凸版印刷，キリンの4社の取り組みについて，企業のトップや担当者自ら執筆し紹介を行っている。そして第Ⅲ部「投稿論文」では，審査を通った4本の論文が収められている（JFBS編集委員会によるdouble-blind review）。それぞれのポイントについては，岡田委員長によるIntroductionを参照されたい。

　今日，持続可能性のもつ社会的・環境的・経済的側面を企業戦略に統合させ考えていくことが，実務家および研究者それぞれの世界において求められており，本書が今後の議論および具体化に向けて一つの材料となれば幸いである。

　JFBSは，2015年9月の大会において「企業家精神とサステナブル・イノベーション」（Entrepreneurship and Sustainable Innovation）をテーマとして議論を行う予定である。持続可能な発展に貢献するイノベーションが，誰によってどこで生み出されるのか，大企業のみならず，中小企業や社会的企業においても検討していくことにする。

　最後に，本書は「企業と社会シリーズ」第4号となるが，創刊号より継続して引き受けていただいている千倉書房，そして担当いただいた同社編集部の神谷竜介氏には，感謝の意を表したい。

2015年5月

企業と社会フォーラム会長／早稲田大学商学学術院商学部教授　谷本　寛治

目　次

はじめに……………………………………………………………谷本寛治… i

Introduction…………………………………………………Masahiro Okada… 1

I　招待論文 /Invited Articles

1　The Stakeholder Route to Successful Sustainability Management
　　………………………………………………… CB Bhattacharya… 15

〈翻訳〉1　サステナビリティ経営を成功させるステイクホルダー・
　　ルート ………………… CB バタチャリア（JFBS 事務局訳） … 44

2　Strategy and Sustainability
　　―Changing Conceptions to Enable a Sustainable Direction
　　……………………………………………………… Nick Barter… 73

3　Corporate Greening
　　―A Conceptual Analysis
　　……………………… Gabriel Eweje, Aymen Sajjad and Mina Sakaki… 95

4　Green Management
　　―Environmental-Financial Performance Nexus and Dimensions of
　　Innovation ………………………………………… Keiko Zaima… 118

5　CSR and Corporate Reputation
　　―Towards Effective Strategy Approaches! ……… Tobias Bielenstein… 136

II　企業ケース /Cases of "CSR and Strategy"

1　持続可能性と積水ハウスの戦略………………………… 和田　勇… 159

2 企業と社会の相互作用としてのステークホルダー・レビュー による
CSR の経営への統合化促進―NEC のケース････ 鈴木　均・遠藤直見…168
3 BtoB 企業の CSR とブランド価値向上
―凸版印刷を事例として………………………………………… 今津秀紀…182
4 キリンの CSV の取り組みについて
―「ブランドを基軸とした経営」による社会と企業の持続的な
発展に向けて………………………………… 太田　健・四居美穂子…195

Ⅲ　投稿論文（査読付）/Reviewed Articles

1 Psychological Empowerment for Competitive Advantage
―A Resource-based Approach to Human Resource Management in Hospital Industry of Nepal
…………… Sunita Bhandari Ghimire and Dhruba Kumar Gautam…207
2 高齢化地域の持続可能性に資する地域企業のイノベーション戦略
―徳島県上勝町「いろどり」からの考察……………………… 芳賀和恵…248
3 韓国の社会的企業の'制度化を通じた'育成に関する考察
―「社会的企業育成法」を巡る現状と課題を中心に………… 金　仁仙…272
4 消費財の情報特性が CSR 活動に与える影響の分析………… 吉田賢一…294

Introduction

Masahiro Okada
Professor, The Graduate School of Business Administration
Keio University, Japan

1. Sustainability and Strategy

This annual book consists of articles examining research questions pertaining to the relationship between "sustainability" and "strategy."

The concept of "sustainability" has never been discussed so frequently and from such a variety of angles as today. The most widespread and traditional use of the term is probably in the sense of sustainability of natural environment and resources of this planet, symbolized in "Spaceship Earth" by Buckminster Fuller in 1963. Based on this context, "sustainable development" which denotes the economic development without sacrificing natural environments and social structures has become the central issue not only in the development sector, but also in societies worldwide including both developing and developed nations.

On the other hand, the concept of "sustainability" has also been at the center of business strategy theories in a totally different sense. Especially in the context of "sustainable competitive advantage" meaning the most desirable outcome of any business strategies, "sustainability" simply denotes "the ability to endure." In strategy theories, how durable thus sustainable a firm's certain competitive position can be is a critically important condition for a firm's competitive advantage. However, this use of sustainability reflects no ecological or social implications of firm strategies. With some important exceptions such as stakeholder theories, relationship theories and social network theories, most mainstream strategy theories such as Porter (1980), Wernerfelt (1984) and Barney (1991) to name a few are all straight "economic" theories with stockholders value maximization as the ultimate dependent variable.

In the business strategy theories mentioned above, a firm's competitive position is to be judged as "sustainable" when either the conditions of 1) path dependence, 2) social complexity, 3) causal ambiguity, or 4) imperfect substitutability is met. However, researchers as well as corporate managers are now facing the fact that "sustainability" of their economic performance is no longer attained by simply optimizing their capabilities and resources to satisfy those conditions. It has become inevitable to integrate social, ecological and economic dimensions of sustainability into firm strategies and related theories. This integration requires innovative ways of management practices, distinctive business models, and new theories and research methodologies.

We hope that this Annual Book provides an opportunity for renewed thinking of the emerging and evolving relationship between sustainability and strategy.

2. Overview of contributions

The papers included in this entire volume show the possibility that the relationship between sustainability and corporate strategy can be discussed and analyzed from multi-dimensional aspects, so that the readers can grasp the more holistic view of this important relationship.

The following five papers in this Part I and four papers in Part II are invited papers, while four contributions in Part III have been selected from presented papers at the JFBS annual conference in September 2014. These selected papers have been reviewed by two referees with double-blinded process.

Contributors in this Part I establish the theoretical and conceptual underpinnings of the relationship between sustainability and strategy. Then in Part II, advanced showcases of notable companies in Japan are presented. In Part III, each contributor discusses the specific research questions relevant to the sustainability-strategy relationship.

2-1. Part I: Invited articles

This part begins with the paper by C. B. Bhattacharya, one of the gurus in this research realm. The author asserts that the firm activities relevant to sustainability not only directly benefits the firm performance through improved efficiency, but also indirectly benefits through the route of stakeholders. By empowering stakeholders by involving them in firm initiatives, three Us" can be enhanced including Understanding of the sustainability programs, the sense of Usefulness, and the Unity among the firm and its multiple stakeholders. This stakeholder route can improve corporate value through building reputation, boosting employee morale, and stronger customer loyalty which then enhances the firm's competitive positions.

The next contribution titled "Strategy and Sustainability: Changing Conceptions to Enable a Sustainable Direction" written by Nick Barter proposes a new way of conceptualizing a link between "sustainability" and "strategy." He asserts that the two should not be treated as separate concepts, but rather be treated as unified entity in order to develop an integrated framework of "sustainable strategies." The background of this conception is the author's firm belief that human being and its action are a subset of natural environment. "Thus, rather than asking about an organization's impact via its products and services on the environment, society or the economy, the strategist needs to ask questions about how the products and services will impact themselves and their associates."

In the third invited paper "Corporate Greening—A Conceptual Analysis" by Gabriel Eweje, Aymen Sajjad, and Mina Sakaki, these authors assume that corporate initiatives to fulfill social responsibility are now shifting from ethical duty to rational decision to be justified on the basis of its own competitiveness. The authors then discuss the concept of "green management", the role of business in mitigating environmental impacts, and concrete examples of initiatives and strategies to reduce companies' environmental impacts such as green-design, life-cycle management, product stewardship, cleaner production, EMS (environmental management systems), and GSCM (green supply chain management).

4   Introduction

The fourth contribution is "Green Management: Environmental-Financial Performance Nexus and Dimensions of Innovation" by Keiko Zaima. This paper gives a brief review of the empirical research on the environmental-financial performance nexus first, and shows that the results are predominantly positive. She suggested three dimensions of innovation in green management, based on a reconsideration of green management concepts and four aspects that a company can reshape through green management.

The fifth contribution is "CSR and Corporate Reputation—Towards Effective Strategy Approaches!" written by Tobias Bielenstein. In this paper, based on a UN report, it is pointed out that most companies fail both to connect sustainability activities to their core business and to global challenges. Then the paper stresses the importance of connecting sustainability activities to corporate reputation, and shows a methodology to develop an effective strategy to maximize "reputational return of CSR." The key is to consider the all nine dimensions of corporate reputation: employer attractiveness, business performance, quality of products & services, innovativeness, marketing & sales effectiveness, management quality, ethical business practice, transparency, and social responsibility.

Overall, these contributions in Part I share the fundamental shift of the nature of sustainability initiatives in common, from ethical mandates for corporations to something to generate positive values to the firm and to strengthen its competitive position.

2-2. Part II: Cases of "CSR and Strategy"

This part consists of four cases of Japanese firms that are effectively engaged in sustainability initiatives to enhance their competitiveness.

First case is Sekisui House, Ltd., a major home builder in Japan, presented by President Isami Wada. The company thinks of the house as the key solution to social problems. Especially the experience of East Japan Earthquake made us realize the house is one of the most important cornerstones of human life. The company holds the "Sustainable Vision" as the backbone of its CSR activities. With this vision, the company has been trying to optimize the balance among four values: 1) economy, 2) environment, 3) society, and 4) people. With this sustainable vision coupled with the concept of "Responsibility for the future", the company is trying to create "shared value."

The second is NEC, a diversified electronics company. In the paper titled "Stakeholder Review as an interaction between the firm and society: Accelerating the Integration of CSR into Corporate Management," Hitoshi Suzuki and Naomi Endo report that the company is now engaged in stakeholder review based on ISO 26000, in which a not-for-profit organization called CSR Review Forum is involved in the periodical meeting with the corporate directors who are in charge of CSR and mid-term strategy plan. The authors also point out the limitations of the review driven by ISO 26000.

The third case "The CSR of BtoB Company and its Brand Management" is written by Hidenori Imazu, Toppan Printing Co., Ltd. The author analyzes the relationship between CSR and brand management in the case of BtoB business by focusing on the role of CSR communication activities. According to the author, the company does not have any principle specifically defines the role of CSR because it is believed that CSR is not separate activities done by a particular department but should be embedded in the philosophy or vision of the entire corporation. The author suggests that the CSR communication strategy should shift from the traditional promotion toward general public in order to improve its reputation, to more strategic communication with investors and professional rating agencies.

The last example is presented by Mr.Takeshi Oota and Ms. Mihoko Yotsui of Kirin Holdings Company, Limited, a diversified beverage company with its beer brewery business as its core, introduces the company's corporate strategy titled "Brand-centered Management." This management is realized through linking the two activities: so-called CSV initiatives and new product development (R&D). Also, the enhancement of organizational capability is taken as a necessary platform on which the brand-centered management will be fulfilled. Then the paper introduces several concrete examples of CSV initiatives including the generation of new alcohol beverages by incorporating the local fruits in Fukushima where the agricultural business was devastated after the East

Japan Earthquake.

2-3. Part III: Reviewed articles

The first paper is "Psychological Empowerment and Competitive Advantage: A Resource-based Approach to Human Resource Management" by Sunita Bhandari Ghimire. The research question of this study lies in the role of human resource empowerment in enhancing competitiveness of the firm. The importance of employee empowerment resonates with the case of Kirin in Part II in terms of the awareness and motivation of employees is the key ingredient of successful execution of intended strategies.

The second paper is "The Innovation Strategy by Small Local Businesses in the Aging Community in Japan" by Kazue Haga. Based on innovation theories and through the case analysis of a small for-profit company called Irodori Co., Ltd. ("various colors" or "colorfulness" in English) in the rural mountainous area of Tokushima Prefecture, the author exemplifies the possibility of innovative model of economic value creation in rural areas. Aged associates of the company picks natural leaves in the mountain and pack them in very-well designed forms, and sell and deliver them to high-class Japanese-style restaurants nationwide. The author concludes that the development process of this company is autopoietic, that the locally embedded low-tech materials can be utilized to enhance the competitiveness of local economy, and that a firm purely designed as a for-profit company can also generate a great deal of social values depending on its context.

The third paper "Social Enterprises in Korea: Social Enterprise Development Act" by In Sun Kim reports the historical development and policy structure of social enterprise development in the country, and identifies problems its government should counteract. This legal scheme for social enterprises emerged as a countermeasure to mitigate the high rate of unemployment right after the Asian Currency Crisis. The paper concludes that the current policy should shift from direct subsidization for labor costs (employment) to indirect measures such as flexible financing, intermediary management consulting services, and enhancement of various types of social capital markets.

The last paper "The Effect of the Informational Nature of Consumer Products on CSR Activities" by Kenichi Yoshida quantitatively verifies the hypothesis that the harder it is for consumers to perceive the quality of the products, the more active the company becomes in its CSR activities. The result shows that especially in the case of "experience durable goods," or experience goods with durability such as automobile, pharmaceuticals and cosmetics, the company tends to be engaged in CSR activities more actively than in other cases.

Throughout these articles, readers may well recognize that social and environmental aspects of business are no longer external factors, but rather indispensable ingredients of successful strategies in the context of Asia.

3. About JFBS

JFBS is a relatively new academy founded in 2011 by Professor Kanji Tanimoto, Waseda University in Japan.

Prior to the formalization as an academy, Professor Tanimoto had initiated the Forum of Business and Society (FBS) in 2009 with twenty eight members as a private research forum. The purpose of the FBS was for academia, industry, non-profit sector, labor and government to cooperate with one another for further advancement of knowledge in the field of business and society relationship. Nine research sessions had been held under the scheme of this FBS.

After the FBS coordinated the 4th annual meeting of the Asia Pacific Academy of Business in Society in Tokyo in 2010, it was determined to make the FBS as an official academy as the Japan Forum of Business and Society (JFBS). Since then, the Academy has been continuously inspiring those who have a keen interest in the evolving relationship between business and society.

4. Conference Outline

The JFBS 4th Annual Conference was held in September 2014, hosted by the Japan Forum of Business and Society (JFBS). The conference provided a platform for the discussion of various dimensions of the sustainability-strategy relationship.

This year's JFBS annual conference was structured in the following manner. The conference was kicked off by two keynote speeches on sustainability and strategy; one from academic side and another from practical side, followed by the first plenary session with three panelists including both keynote speakers. Then the conference went into the series of breakout sessions.

Breakout sessions were conducted under eight themes: 1) Sustainability and Brand Management, 2) Sustainable Strategy and Emerging Markets, 3) Sustainability and Capital Markets, 4) Green Management, 5) Development and Strategy, 6) Sustainability and Strategy, 7) International Engagement, 8) Sustainability and Competitive Advantage, and 9) CSR and Stakeholder. The first four themes were discussed as organized sessions, and the latter five were discussed based on submitted research papers.

Finally, the Plenary Session 2 was designed to summarize the entire conference with each session chair briefly reporting the content of discussion, and all participants exchanged opinions to synthesize the learning through various sessions.

One other trial was made in this year's conference. We allowed presentations with different languages, namely English and Japanese, together in the same session, prioritizing the similarity and synergy between the contents of presentations, rather than pursuing practical efficiency of grouping presentations in the same language.

The outline of the sessions are summarized below. As for the detail of this entire conference, please refer to the following website:
http://j-fbs.jp/annualconf_2014_en.html.

4-1. Keynote Speech: "Sustainability and Strategy" by C.B. Bhattacharya, European

School of Management and Technology, Germany, and Isami Wada, Sekisui House, Japan.

4-2. Plenary session 1
"Sustainability and Strategy" chaired by Masahiro Okada, Keio University, Japan, with panelists: Nick Barter, Griffith University, Australia, C.B. Bhattacharya, European School of Management and Technology, Germany, and Isami Wada, Sekisui House, Japan.

4-3. Organized breakout sessions
4-3-1. "Sustainability and Brand Management" chaired by Tobias Bielenstein, Branding-Institute, Germany, with panelists: David Hessekiel, Cause Marketing Forum, USA, Hidenori Imazu, Toppan Printing, Japan, and Takeshi Ohta, Kirin, Japan.
4-3-2. "Sustainable Strategy and Emerging Markets" chaired by Masahiro Okada, Keio University, Japan, with panelists: Ashir Ahmed, Kyushu University, Japan, Yoko Nagashima, Hewlett-Packard Japan, Japan, and Ahok Roy, JED, Japan.
4-3-3. "Sustainability and Capital Markets" chaired by Hiroshi Amemiya, Corporate Citizenship Japan, Japan, with panelists: Tsukasa Kanai, Sumitomo Mitsui Trust Bank, Japan, Miho Kurosaki, Bloomberg, Japan, Toshiaki Yamamoto, Osaka Electro-Communication University, Japan, and Masahiro Yokoyama, Daiwa Securities Group, Japan.
4-3-4. "Green Management" chaired by Gabriel Eweje, Massey University, New Zealand, with panelists: Takenobu Shiina, Suntory Holdings, Japan, Maho Takahashi, Lush Japan, Japan, Keiko Zaima, Kyoto Sangyo University, Japan.

4-4. Breakout sessions based on call for papers
4-4-1. "Development and Strategy" chaired by Kazuyori Kanai, Osaka University of Commerce, Japan, with presentations "The Analysis of the Use of Capital Markets to Develop Social Enterprises in Korea" by Insun Kim, Beijing University, China, and

"The Sustainability and Strategy of Aging Local Communities: A Case of Kamikatsu-cho, Tokushima, Japan" by Kazue Haga, DIJ (Germany Institute for Japan Studies) Tokyo, Japan.

4-4-2. "Sustainability and Strategy 1" chaired by Masao Seki, Sompo Japan Nipponkoa Insurance, Japan with presentations: "The Competitive Advantage of Small and Medium Sized Enterprises through Sustainability" by Masaki Wada, Shigeru Machii, Hisashi Oka, and Yukako Kunida, Global Sustainable Business Alliance, Japan, and "'Dynamic Equilibrium' and 'Co-creation' between Business and Society toward Sustainability—From the Viewpoint of Field Theory" by Hideya Nagai, Toyo Gakuen University, Japan and "The Dissemination Process of Social Innovation and its Evaluation: the Case of 10 Yen Donation Per One Package Project by Yamato Group" by Masaatsu Doi, Hosei University, Japan and Yuki Misui, Takasaki City University of Economics, Japan.

4-4-3. "Sustainability and Strategy 2" chaired by Makoto Fujita, Waseda University, Japan, with presentations: "Public Goods Provision as a Result of Implementing Sustainability Strategies — The Result-oriented Perspective as a Tool in Stakeholder-guided Strategy Development" by Tobias Bielenstein, Branding-Institute, Germany, "Recent Development of Strategy Theories—Unification of Business Strategy and Social Problem Resolution" by Jun Oeki, Tokyo University of Science, Japan, and "A Study on CSR and CSV" by Hidemi Tomita, Lloyd's Register Quality Assurance Japan, Japan.

4-4-4. "International Engagement" chaired by Philippe Debroux, Soka University, Japan, with a presentation: "Ethical Capital as Strategic Resource for Chinese Inclusive Business" by Po-Keung Ip, National Central University, Taiwan.

4-4-5. "Sustainability and Competitive Advantage" chaired by Nick Barter, Griffith University, Australia, with presentations: "Platform-Integrated Cap and Trade with Renewable Energy Technologies" by Grace T.R. Lin, Jen-Sheng Wang, and Po-Hsin Hsieh, National Chiao Tung University, Taiwan, "Rate the CSR Raters in China" by Hsiang-Lin Chih, National Taipei University, Taiwan, and Mei Sun, Sichuang University, China, and "The Strategy of Vale S.A. for Urban Infrastructure Shortage Reduction: Strategic Relationship Behavior with Local Governments as a Competitive Advantage for Sustainable Business" by Felipe Francisco De Souza, Masaru Yarime, Andre Sorensen, Isabel Ache Pillar, and Andreia Rabetim, The University of Tokyo, Japan.

4-4-6. "CSR and Stakeholder" chaired by Shinji Horiguchi, Kobe University, Japan, with presentations: "The Current Status and Limitations of Multi-Stakeholder Perspective in the Theories of the Firm" by Nobuyoshi Omuro, Kyoto Sangyo University, Japan, and "Stakeholder Review as an Interaction between the Firm and Society: Accelerating the integration of CSR into Corporate Management" by Hitoshi Suzuki, Institute for International Socio-Economic Studies, Japan, and Noami Endo, NEC, Japan.

4-4-7. Doctoral Session chaired by Philippe Debroux, Soka University, Japan, Masahiro Okada, Keio University, Japan, and Kanji Tanimoto, Waseda University, Japan. Presentations: "The Evolution of Corporate Governance System and CSR: A Case Study of the Growth Strategy of Volks Wagen" by Naomi Yamaguchi, Hitotsubashi University, Japan, "Determining Factors of CSR Activities by Japanese Corporations: The Effect of Informtional Nature of Consumer Goods on CSR Activities" by Kenichi Yoshida, Waseda University, Japan, and "The Effectiveness of Issue Selling against the Organization of Donation Platform" by Natsuko Matsuno, Waseda University, Japan.

4–5. Plenary session 2

With all the sessions done, the whole participants of the conference reassembled and had a discussion to synthesize what we learned in this final plenary. Chair persons of all the sessions lined up and made a brief summary of each session. One of the several conclusions we reached there was the importance of education, not only for students but also for those who are in business communities in terms of the holistic nature of business phenomena which include environmental, social and economic spheres. Other lessons included the necessity of a new scheme to evaluate corporate performance in light of multifaceted nature of firm activities.

5. JFBS Annual Conference 2015

JFBS will hold the 5th annual conference on September 2015 under the theme of Entrepreneurship and Sustainable Innovation. Sustainable Development is defined as "development that meets the needs of the present without compromising the ability of future generations to meet their own needs" and has been a main theme in the last two decades. Now we need a new step to promote sustainable development in the next decade. Social entrepreneurship is expected to tackle challenges for sustainable development and take the sustainability initiative. Sustainable/social innovation creates new social values through businesses which tackle social and environmental problems with a view to their resolution. To create a new innovation, we need to build and promote collaboration among social business, big business, NGO, local people, government, and academia, and to make a institutional environment to enhance new businesses.

In the coming conference, the issues we will discuss include, but not limited to, social entrepreneurship, sustainable/social innovation, innovation management, partnership among sectors, regional development, sustainable consumption and production, and social marketing.

<References>
Barney, J. B. (1991) 'Firm Resources and Sustained Competitive Advantage', *Journal of Management*, Vol. 17, No. 1, pp. 99-120.
Porter, M. E. (1980) *Competitive Strategy*, Free Press: New York.
Wernerfelt, B. (1984) 'A resource-based view of the firm', *Strategic Management Journal*, Vol.5, Issue 2, pp. 171-180.

I 招待論文 /Invited Articles

1 The Stakeholder Route to Successful Sustainability
   Management　　　　　　　　　　　　　　CB Bhattacharya
   〈翻訳〉1　サステナビリティ経営を成功させる
   　　　　ステイクホルダー・ルート　CB バタチャリア（JFBS 事務局訳）
2 Strategy and Sustainability
   ―Changing Conceptions to Enable a Sustainable Direction
   　　　　　　　　　　　　　　　　　　　　　　　Nick Barter
3 Corporate Greening ―A Conceptual Analysis
   　　　　　　　　　Gabriel Eweje, Aymen Sajjad and Mina Sakaki
4 Green Management
   ―Environmental-Financial Performance Nexus
   　and Dimensions of Innovation　　　　　　　　Keiko Zaima
5 CSR and Corporate Reputation
   ―Towards Effective Strategy Approaches!　Tobias Bielenstein

1 The Stakeholder Route to Successful Sustainability Management

CB Bhattacharya

Professor, European School of Management and Technology in Berlin, Germany

[Abstract]
The following contribution contains a three-fold focus on stakeholders. First, I will describe how companies can empower stakeholders so that they become co-creators of sustainability–a process that draws them closer to the activities for better Understanding, ensures that programs are optimally Useful, and enhances Unity, the stakeholders' sense that the company shares their values. Second, I will point out the important role of communication; that is, how companies can get their message out in ways that draw stakeholders closer to the company. Third, I will provide guidance on how to quantify key components and discover linkages between them so that companies can calibrate their sustainability activities on a long-term basis.

1. Introduction

Company performance can be greatly enhanced by successful management of sustainability. Some of this performance is driven by the direct savings of sustainability activities themselves. Reductions in carbon emissions may lead to savings on energy bills. Decreasing and recycling waste from production may lessen the need to purchase materials from suppliers. These savings may be thought of as the direct route through which sustainability provides value for a company. The direct route is often tangible and relatively straightforward to measure.

However, what can also generate value for companies is a secondary and complementary route through sustainability initiatives. In this route—which we call the stakeholder route-the value from sustainability stems not from cost savings inherent in the initiative, but rather from the reactions of stakeholders when they learn about the sustainability initiative(s). Our research, described in more detail in *Leveraging Corporate Responsibility: The Stakeholder Route to Maximizing Business and Social Value* examines this alternative pathway to corporate performance, a pathway that is complex, often quite intangible, and sometimes challenging to measure.

Companies that are able to unlock its potential and build strong stakeholder relationships stand to benefit greatly in both the near and long term. Surveys of senior executives and sustainability professionals indicate that this route can generate corporate value in a number of ways—such as, building reputation, enhancing employee morale, and strengthening competitive positions (Bonini et al., 2009). A company's positive record of sustainability fosters consumer loyalty and, in some cases, can turn customers into brand ambassadors and advocates who may be willing to even pay a price premium to support the company's social policies (Sen and Bhattacharya, 2001). Sustainability may offer a competitive advantage in attracting, motivating, and retaining talented employees (Greening and Turban, 2000) and socially responsible investors (Sen, Bhattacharya and Korschun, 2006; Hill et al., 2007).

It is to be expected that benefits that stem from the stakeholder route depend on a company's ability to understand and respond to the needs and wants of its stakeholder base. We call this **stakeholder centricity**. Companies that remain attuned to customers, employees, investors, local communities, and others stand to gain substantial rewards from their sustainability initiatives; in contrast, companies that ignore potential stakeholder reactions to sustainability may suffer substantial repercussions if those activities damage stakeholder relationships. In this article, we examine three critical ways that stakeholder centric companies leverage sustainability as they cultivate strong relationships with stakeholders. First, we will show how companies can empower stakeholders so that they become co-creators of sustainability. Second, we will point out the important role of communication; that is, how companies can get their message out in ways that draw stakeholders closer to the company. Third, we will provide guidance on how to quantify key components and discover linkages between them so that companies can calibrate their sustainability activities on a long-term basis.

2. Co-creation of the sustainability strategy

We promote a co-creation approach to sustainability strategy. In fact, it is hard for us to imagine a truly stakeholder-centric organization that does not seek stakeholder engagement from stakeholders as it creates its sustainability strategy. Properly involving stakeholders in formulating corporate responsibility programs result in more focus and success. Co-creation allows stakeholders to become part of the solution and reduces the gap between their expectations and the firm's response. To do this, companies must first understand the overarching factors that make co-creation of sustainability successful, and then go through a process by which stakeholders can fully co-create sustainability.

2-1. Success factors of co-creation strategies

Our research has proven that companies have been empowering their employees and consumers as co-creators of effective sustainability strategies by putting them at the centre of their sustainability efforts. For example, matching gift programs, in which a company matches employee donations to charities dollar-for-dollar, are popular with employees, because the employees can individually decide what cause they want to support (e.g., environmental organization, community sustainability) and how much they want to contribute. However it is not always clear what factors are of primary importance when seeking to successfully implement sustainability initiatives.

Based on both our research and our conversations with executives, we have identified three factors that drive the success of the CR implementation strategy:

1)Whether the stakeholder is an enactor or merely an enabler;

2)Whether the connection with others in the company and its program is primarily horizontal or vertical;

3)Whether management keeps the process of evaluating the program formal or informal.

These factors can guide managers as they implement sustainability strategies. Overall, our research suggests that in order for companies to design sustainability initiatives in stakeholder-centric ways, managers need to put formal processes in place to foster informal exchanges between stakeholders. Such formal systems will not only provide stakeholders with the rewarding experiences that they are looking, for but may also serve as a marketing research tool for managers who can use feedback from these processes to learn more about stakeholder needs and preferences. Such formal processes could be to define stakeholder roles and tasks, to integrate these into company operations, and to obtain stakeholder feedback. Thus, we recommend that processes include organizational monitoring, evaluation, and reward systems.

2–2. Facilitating better connections

A stakeholder-centric and co-creation-oriented approach facilitates dialogue between a company and its stakeholders and incorporates this input as early as possible in the sustainability formulation process. A critical step in co-creation is to examine how to enhance the benefits that stakeholders will get from sustainability activities—with a special eye on involving stakeholders in the process of co-creating a company's sustainability initiatives to maximize all stakeholders' advantages. This way, companies can not only support the creation of additional value for stakeholders through the co-creation process itself, but also guarantee that a full range of relevant programs and issues are considered. The relevant stakeholder groups have created whatever is chosen in the end.

How are managers to put these insights into action? There are four steps to successful sustainability formulation: articulation, generation, distillation and selection. Articulation and distillation are more reliant on managerial input, while generation and selection present co-creation opportunities, actively engaging key stakeholders in sustainability decision making.

2-2-1. Articulation

The decision of what type of sustainability value should be created and formal articulation of objectives of sustainability is the most crucial task for those engaged in formulating sustainability strategy. Only after this step can one decide on the most appropriate stakeholder groups to target for involvement. The end should be clear, or it will be extremely difficult to determine the means to achieve sustainability value. Sustainability strategy may differ substantially according to the desired goal. A much broader stakeholder focus is needed when the goal is to improve corporate reputation and brand equity than when the goal is to attract and retain employees.

Those formulating sustainability strategy need to determine the exact objectives and articulate these in such a way as to create buy-in. Corporate leadership has to offer initial support of the initiative so that the necessary resources are available and continue to support the initiative long-term to ensure implementation and results. A valuable way to better co-create CR goals with fellow employees and get broad-based internal buy-in is to build shared understanding among colleagues by learning from knowledgeable others in the business.

2-2-2. Generation

Following the identification of relevant stakeholders and determination of objectives we recommend that companies engage and interact with stakeholders to generate a portfolio of sustainability initiatives that are in line with the objectives. Companies can do this by engaging individual stakeholders in qualitative research, "digging deeper" into their personal needs, whether tangible or more psychological.

Our research has shown two overarching psychological benefits for a stakeholder when he or she is affiliated with a socially responsible company, namely self-esteem and self-coherence. Sustainability can improve a customer's self-worth, making him feel good about buying from a sustainable company, and it can increase his self-coherence by demonstrating that the company has similar values to him, enabling him to feel that he has continuity across his different life spheres (i.e., work, leisure, personal).

Overall, understanding the needs of the stakeholder—the essence of stakeholder centricity—can help managers hone in on features that are best-suited for co-creation.

2-2-3. Distillation

After a set of CR initiatives have been generated in collaboration with stakeholders, managers still have to prioritize or "distil" the initiatives, putting them in line with the core competencies of the company and what the company can best deliver. Even the best sustainability initiative is wasted effort if the company cannot correctly implement it.

A practical approach to the distillation phase is to create a matrix (often called a materiality matrix) plotted according to the dimensions: the importance or attractiveness to stakeholders and the importance to the company in terms of the likely influence of the initiative(s) on business success. This phase should filter the co-created set of alternatives from the generation phase to a smaller set that can be implemented successfully by the company.

2-2-4. Selection

Subsequent to the steps of articulation, generation, and distillation, there should be a relatively broad understanding of the parameters of the initiative that will be selected and implemented. Managers can now select the top initiatives and finalize which stakeholder segments to target for each initiative in the final selection stage. A stakeholder-centric company will use stakeholder's understanding and attitudes about initiatives as a primary (but not the only) driver of selection. In what might be thought of as an additional piece of stakeholder input, stakeholder surveys can be conducted to quantify the value that various initiatives are likely to generate. This data can inform the managers' final decisions about which initiative to implement.

Within the selection stage, those deciding upon sustainability initiatives within the organization may also begin to partner with a non-profit organization (or organizations) to implement the initiative. A development team, with members from the company and the NGO, can describe the sustainability activity as a group of features such as the cause, the implementation of the cause (marketing, volunteering, etc.), proposed non-profit partners, proposed type and level of stakeholder engagement. Companies and NGOs should begin early to find a "common ground" to build long-term partnerships; this will involve co-creation with the NGO, another stakeholder.

3. Communication of sustainability strategy

We explained how companies can become more stakeholder centric through creation of sustainability initiatives with stakeholders in the previous section, Such co-creation can enhance the benefits that stakeholders derive from those initiatives. However, stakeholder-centricity also involves making sure that stakeholders have a thorough understanding of the company's sustainability efforts so that they derive benefits and reward the company in kind. We now examine strategies for communicating sustainability initiatives so that stakeholders develop strong and lasting relationships with the company. Successful strategies address the why, what, and where of communication, as well as take the particular circumstances of the company and its stakeholders into account.

3-1. Key challenges of communication

Since communication is often the primary means through which stakeholders develop an understanding of sustainability programs, it is has to be a critical component of sustainability strategy. This understanding becomes the starting point for stakeholders to interpret sustainability activities and to determine whether they wish to deepen or sever ties with the company. Since few stakeholders are able to participate directly in all sustainability initiatives the role of communication is to ensure that stakeholders are at least able to learn about those initiatives which are most likely to produce value for the company.

Managers seeking to communicate sustainability effectively face two formidable challenges. Awareness of individual initiatives is often low, and while this is changing, many stakeholders continue to be quite sceptical of sustainability programs to begin with.

・**Stakeholder awareness.**

Needless to say, sustainability cannot produce business value through the stakeholder route until stakeholders become at least aware the company's commitment to sustainable practices. Awareness can thus be thought of as a necessary but not sufficient outcome of sustainability communication. However, our research reveals that awareness of a company's social and environmental sustainability practices is quite low. In one of our studies, only 16% of consumers were aware. In another of our studies, a nationwide survey of employees, we found somewhat higher levels, but still only about 50% of employees were aware of their companies' social or environmental initiatives. Consistent with our findings, of the 20 attributes measured in the annual Harris Interactive corporate reputation study, people are mostly in the dark about corporate responsibility: questions about whether companies are socially and environmentally responsible consistently elicit the most "don't know" responses (The Harris Poll Reputation Quotient, 2012). This makes awareness a key stumbling block in the company's quest to reap strategic benefits from a social responsibility program (Sen et al., 2006; Du et al., 2007; Bhattacharya et al., 2008).

・**Stakeholder scepticism.**

Many stakeholders are eager to learn about the sustainability strategies of companies with which they interact. But this eagerness should not be mistaken for naïveté. Many people are bombarded with media messages about the misdeeds of companies, and "green washing", where companies exaggerate their sustainability accomplishments (Sen et al., 2009), has left many people jaded. A formidable challenge for managers wishing to communicate sustainability communication is to confront, with the goal of minimizing, the scepticism that many stakeholders have.

The scholarly literature, including our own research, indicates that when stakeholders learn about sustainability initiatives, they often make judgments about what is motivating the company: a genuine concern for social and environmental welfare or a profit seeking motive. Communications need to address these attributions of the company's motivations (Yoon et al., 2006). However, we do not advocate taking the approach that some companies do: keeping a low sustainability profile and hiding the benefits that the company gleans from those efforts. This is because recent research on sustainability attributions suggests that stakeholders are often tolerant of profit-driven motives as long as sustainability engagement is also attributed to motives of genuine concern (Sen et al., 2006). In fact, some stakeholders actually applaud sustainability initiatives that make both environmental and business sense and reward these companies for these sorts of innovations.

3-2. Message content

The focus of typical sustainability communication is a company's involvement in social causes. There are several factors a company can emphasize in its sustainability communication, such as its commitment to a cause, the impact it has made on the cause, and the congruity between the cause and the company's business (i.e., sustainability fit).

• **Sustainability commitment.**

An important first step in communicating sustainability is demonstrating that the company has made a commitment to the cause. Typically, stakeholders will evaluate a company's commitment based on three aspects:

- the amount of input (e.g., dollars donated, in-kind contributions, corporate resources such as marketing expertise or human capital)
- the consistency of input (Dwyer et al., 1987)
- the duration of the commitment

Of these, perhaps the most underappreciated by managers is the durability of the commitment. Because long-term commitments are difficult to fabricate, they tend to be more credible. This idea is supported by research showing that longer term commitments were more likely to be seen as driven by a genuine concern for increasing societal/community welfare than shorter term campaigns.

・**Sustainability impact.**

Stakeholders not only like to see that a company is strongly committed to sustainability; they also look for evidence that the initiatives are actually having an effect on societal welfare. Managers can achieve this by focusing on sustainability outputs as much as inputs. Research has documented positive associations between perceived societal impact of a company's sustainability initiative and consumers' attributions of genuine concern, resulting in consumers' advocacy behaviours toward the company (Du et al., 2007).

・**Sustainability fit.**

Another important factor to communicate is sustainability fit, or the perceived congruence between a sustainability issue and the company's business. Stakeholders expect companies to sponsor issues that have a logical association with the company's core activities (Cone, 2007; Haley, 1996). Sustainability fit may result from product features (e.g., a herbal products brand sponsors the protection of rain forests), affinity with specific target segments (e.g., Avon fights breast cancer), or corporate image associations created by the brand's past conduct in a specific social domain (e.g., Ben & Jerry's and the Body Shop's activities in environment protection.) (Menon and Kahn, 2003).

CSR fit is important because it affects stakeholders' attributions (Menon and Kahn, 2003; Simmons and Becker-Olsen, 2006).When there is no logical connection between sustainability efforts and a company's core business, stakeholders often become sceptical as they attempt to figure out what the company's true motives are. Higher fit sustainability initiatives are more consistent with stakeholders' understanding of the company and hence, tend to be attributed to genuine concern for the cause. But even initiatives that seem low fit at first blush can be framed so that stakeholders are more accepting.

3-3. Communication channels

Information about a company's sustainability activities or record can be disseminated through various communication channgels. In general, companies can rely on company controlled channels or external channels to communicate their sustainability initiatives. Determining the precise mix involves trade-offs which we describe below.

- **Company channels.**

A great deal of sustainability reporting is generated by the company itself. This may occur through a number of channels depending on which stakeholder audience needs to be reached. Some common forms of company controlled channels are:
- Sustainability reports
- Websites
- Email announcements
- Advertising
- Press releases
- Product packaging

According to an extensive analysis conducted by KPMG in 2008, nearly 80 percent of the largest 250 companies worldwide issued corporate responsibility reports, up from about 50 percent in 2005 (KPMG, 2008). Diet Coke has been running TV advertisements to help raise women's awareness about heart disease, and the brand has also set up a website (www.dietcoke.com/reddress) to communicate the brand's commitment to the cause. Similarly, Stonyfield Farm prints messages on the lids of its 6oz cup yogurt to communicate the company's involvement in a wide variety of health and environmental initiatives to stakeholders.

Company controlled channels are attractive for many companies because, as the name suggests, they enable the company to speak with a single voice that is highly consistent with corporate and sustainability strategy. As the name suggests, these channels give the company considerable control over the messages that reach stakeholders; commitment, impact, and fit can be optimally communicated, which may better address the particular needs of any stakeholder audience. The downside of such channels is that they may be less believable than other communications; research shows that consumers react more positively and less sceptically to information that is deemed to be from an independent source (Yoon et al., 2006; Simmons and Becker-Olsen, 2006). Thus a message that a stakeholder receives through an internal channel may be less successful at overcoming scepticism than a message received from a third party.

・**External channels.**

A counterpoint to such company-controlled communication channels is the large and increasing number of external channels of communication. The media, customers, monitoring groups, consumer forums/blogs are just a few examples of external groups that are capable of communicating to wide swaths of stakeholders. Of course, these influencers are not under direct control of the company, so their messages may not be in-line with the company's sustainability strategy. However, due to their independent nature, stakeholders are likely to trust the information they receive from these external channels more than information they receive from the company. It is incumbent upon companies to maintain a dialog with these groups and individuals so that externally generated messages are as consistent with the goals of the company as possible.

With the continual evolution of social media, almost anyone can become an external generator of content about the company. Consumers may post opinions on Facebook or "retweet" corporate messages to others. Companies like Stonyfield Farm and Ben and Jerry's have been benefiting from consumer ambassadors who raved, in the virtual world, about their social responsibility endeavours. And employees typically have a wide reach among other stakeholder groups through their social ties and are often considered as a credible information source (Dawkins, 2004). As a result, leading companies engage in virtual dialogs with stakeholders in order to rally them around sustainability initiatives and to manage expectations about the company's sustainability performance (Korschun and Du, 2013).

3–4. Communication effectiveness and its moderators

Like all well-managed communications, sustainability communications should be designed specific to a company. Leading companies tailor their communications to the unique situation in which sustainability initiatives are communicated. We find that the two most important characteristics for managers to consider are those of the company and of the stakeholder audience.

- **Company specific factors.**

We find that sustainability communications benefits some companies more than others. For example, companies with favourable reputations are at a relative advantage in terms of stakeholder reactions because often serves as a pre-existing schema upon which stakeholders rely to interpret information about the company (Fombrun and Shanley, 1990). Companies with strong reputations are perceived to have high-source credibility, and thus the positive effects of sustainability communications will be stronger for these companies. In contrast, the effects of sustainability communications are dampened and can even backfire for companies with poor reputations (Yoon et al., 2006).

More specifically, companies that have a long-standing reputation as sustainability leaders (e.g., Timberland, Ben and Jerry's) may find that the effects of sustainability communications are amplified. For example, in the U.S. supermarket sector, the Whole Foods Market brand is strongly positioned on CSR, as it espouses the core value of "caring about our communities and our environment" (Whole Foods Market, 2012). Moreover, this value pervades virtually every aspect of the Whole Foods' business, such as organic and sustainable sourcing to environmentally sensitive retailing, given that the company devotes at least 5 percent of its annual profits to a variety of causes and encourages its employees to engage in community service during working hours.

- **Stakeholder specific factors.**

Just as sustainability communications must be tailored to the company characteristics, so too must communications take into account the unique traits and preferences of the stakeholder audience. For example, NGOs tend to seek hard evidence of the social impact of a company's CSR programs. In contrast, socially responsible investors may be equally concerned with the business impact of sustainability initiatives: how they are linked to key business metrics such as customer equity, employee retention, corporate governance, and risk management. In addition, professional audiences may require adherence to leading reporting standards such as the Global Reporting Initiative (GRI) and AccountAbility's AA1000, which lay audiences may be more concerned with making an emotional connection to the company.

Even within these broader types of stakeholders (i.e., NGO's versus investors, etc.), individuals have personal preferences about which causes are dear to the heart. Sustainability communications are most effective when they address causes that a stakeholder cares greatly about, a phenomenon that is consistent with much research that finds that people pay greater attention to information perceived as self-relevant (vs. non-relevant) (Petty et al., 1981). Self-relevant communications are best-suited to break the media clutter and receive support from stakeholders. For example, in a recent cause promotion by GAP, the U.S. apparel company, the company donated 5 percent of each dollar amount spent at GAP to one of six non-profit organizations directly chosen by the consumer from a list. The six non-profit organizations were selected by GAP based on the wide range of social issues they were collectively representing—from domestic issues such as education (Teach for America) and child hunger in the U.S. (Feeding America), to global issues such as the environment (World Wildlife Fund) and diseases including AIDS, TB, and Malaria in Africa (the Global Fund). By allowing stakeholders to choose what issue and what non-profit organization to donate to, GAP was able to enjoy high issue support from its consumers and thereby enhance the effectiveness of its CSR communication.

4. How to calibrate sustainability strategy

Until quite recently, most consumers, employees, and investors would not integrate sustainability into their decisions about whether to work with a company. Today the landscape looks very different. To stay current with stakeholders' needs and preferences, stakeholder-centric companies continually evolve alongside stakeholders, calibrating sustainability strategy along the way.

How can companies stay current? This section provides the capstone to successful sustainability management: calibrating sustainability strategy. The following pages out-

1 The Stakeholder Route to Successful Sustainability Management   29

line techniques for quantifying value and ways to turn these measurements into actionable knowledge that can be disseminated throughout the company. Such calibration involves three managerial processes:

- Quantifying stakeholder perceptions and business value
- Finding linkages between sustainability and business value
- Identifying opportunities to maximize business value

4-1. How to quantify stakeholder perceptions and business value

The familiar adage "You can't manage what you can't measure" is as true in the sustainability realm as it is in other business domains. If managers are to maximize returns to sustainability investment, then stakeholder reactions—both behavioural responses and psychological interpretations—must be measured.

We rely upon a number of well proven methods and techniques in quantification of stakeholder reactions to CR. These involve both surveys and experiments conducted in the field as well as in the lab. The research also involves a varied set of stakeholder segments such as employees, consumers, and business-to-business customers. But the common theme is clear: any aspect of stakeholder relationships that may substantially affect corporate value should be measured. This approach means answering two important questions:

- How are stakeholders currently interpreting the company's sustainability activities?
- What are the desired outcomes that the company seeks through the stakeholder route?

4-1-1. Current stakeholder reactions to sustainability

Optimization of sustainability engagement requires managers to have reliable means of getting into the mind of the stakeholder in order to develop a sound understanding of the operation of the psychological process. We find that the psychological process behind stakeholder reactions clusters around three core concepts: their Understanding of the initiatives, the degree to which they find the initiatives to be personally Useful, and the Unity they feel with the company. The strongest and most lasting stakeholder relationships are founded on these three concepts, which, when working harmoniously, can result in substantial rewards in terms of corporate performance.

・**Understanding.**

In its simplest terms, Understanding is the collection of perceptions that a stakeholder holds about the company's sustainability. Understanding develops as stakeholders learn about the company's sustainability, and as they ask questions such as: What is motivating the company to become more sustainably? Is this sustainability initiative benefiting society in the way that the company says it is?

Readers may note that the above questions reflect the notions of attributions and of sustainability impact that we described in the section on sustainability communication. We advise companies to measure attributions both of genuine concern and also of profit motivation. Stakeholders tend to balance these two attributions in their interpretations of CR, and since stakeholders sometimes reward companies precisely for the way in which they combine these two motivations—because they see that approach as innovative—it is best to measure both types so that this interaction can be teased out. Another component of Understanding is the sustainability initiative's perceived impact—which should be measured at two levels: the program level and the company level. Measuring this impact of a particular program is a matter of asking about Understanding with questions that draw out whether and how the programs are "making a difference." Measuring the overall efficacy of a company's CR program involves comparing the company to other companies in the same perceived peer group, which is a stakeholder's "baseline" perception of sustainability impact.

・**Usefulness.**

Stakeholders can derive many types of benefits, posing a challenge to managers wishing to measure the extent to which stakeholders find sustainability to be Useful. To gain a holistic sense of Usefulness, managers need to look at both functional and more psychological benefits. Functional benefits are those that are largely tangible and are the direct result of features of sustainability initiatives. For example, in one of our studies involving an oral care program, parents reported that the program resulted in their children having clean and healthy teeth. Another example of a functional benefit is the energy savings to consumers after purchase of CFL light bulbs.

Often in the stakeholder route to sustainability value, benefits are more psychological, stemming from the identity needs of stakeholders to maintain self-esteem and a coherent sense of self. These needs are common to all types of stakeholders, no matter what the nature of their relation with the company may be; however, they often manifest themselves in ways that are specific to a situation or role, requiring managers to measure them with a diverse set of contexts in mind. For example, self-esteem can be very low in the work context even though it is high in the context of one's family life. In such a situation, we advise using questions that refer to self-esteem at work or self-esteem at home (or both).

- **Unity.**

At the core of stakeholder-centricity is the notion of a stakeholder's sense of Unity with the company, a summative concept that reflects the overall quality of the stakeholder company relationship. Unity indicates whether the stakeholder thinks the company shares his or her values, and whether it is trustworthy enough to warrant continuing the relationship into the future. Typical marketing measures such as customer satisfaction may provide a window into Unity, but they are insufficient at helping us understand the true bonds that can form between individuals and companies, because they do not capture the underlying evaluations that stakeholders make about a company's character or "soul."

We measure Unity in two ways. The first is organizational identification, which is the extent to which a stakeholder feels a sense of oneness with, or belongingness to, the company. Researchers Bergami and Bagozzi (2010) have empirically tested and validated a two-item scale that is both reliable and extremely easy to implement. It allows respondents to indicate the overlap between themselves and the company in two ways: verbal and pictorial. One item in their two-item scale asks the respondent to indicate the degree to which they agree or disagree with the statement, "The values of this company match my own values." The other item is a pictorial scale that uses a series of Venn diagrams involving two circles with varying degrees of overlap; the respondents simply choose the picture they perceive to best represent the overlap between themselves and the company. Combining these two scores yields a meaningful measure of the stakeholders' perceived match between themselves and the company.

Not every stakeholder, in every corporate context, however, will feel so close to the company that they perceive an overlap in values. However, Unity may also manifest itself as a heightened sense of trust, a stakeholder's confidence in a company's reliability and integrity (Morgan and Hunt, 1994). Trust is integral to any stakeholder's reaction to sustainability, and is more sensitive to exposure to sustainability information than organizational identification, which tends to be more stable over time. Therefore, it is advisable to measure both identification and trust when quantifying Unity.

4-1-2. Desired outcomes of the stakeholder route

We have argued that much of the value that sustainability initiatives generates is derived from stakeholder reactions. In evaluating sustainability initiatives, it is essential to quantify this value, and when possible, even put a dollar value on stakeholder behaviours. For example, a manager may expect improvements in customer loyalty as a result of sustainability and set a goal of reaching the level of a market leader in its industry. Ultimately, what matters most is not how the desired outcomes chosen—which will differ by company and by management style—but that there is a clearly articulated goal of encouraging stakeholders' behaviour in some way through sustainability activity.

The most useful way to measure sustainability value is to track the behaviours of individual people who interact with the company as consumers, employees, investors, or in other capacities. Simply citing correlations between overall sustainability activity and changes in aggregate performance is not terribly helpful for managers engaged in the complex day-today challenges of sustainability decision-making. Basing decisions on a tenuous link between, say, overall sustainability spending and corporate performance can be misleading at best and dangerous at worst, because it cannot account for the idiosyncrasies in stakeholder behaviour. Of course, these individual behaviours will end up aggregating to improve performance, but measuring them only at the aggregate level inhibits a researcher from knowing which stakeholder segments responded favourably and which did not. Thus, we advise choosing a few important behaviours that can be observed for someone in a stakeholder segment. For example, consumer purchase has tangible value for the company and can be observed objectively (when it is impractical to measure actual behaviours, managers may measure behavioural intentions such as purchase intent instead). Likewise, a person who applies for a job provides another such behavioural response.

An important insight in the framework is that sustainability value can involve societal as well as business value. Therefore, we recommend tracking stakeholder behaviours towards the selected sustainability cause whenever possible. These societal behaviours should be considered as an important component in the case for engaging in sustainability. How does a researcher or manager go about measuring such responses? Societal value of this kind can be measured through surveys and interviews with stakeholders by asking whether and how the company's sustainability has encouraged them to increase their donations, volunteering, or positive word-of-mouth for charities that address the same social issues as the company's sustainability programs. A fine example of these sorts of measures is a study that examined and measured consumer donations to charities sponsored by the company from which they purchase products. In that study, the more consumers felt a sense of Unity with the company, the more they reported donating to these charities (Lichtenstein et al., 2004).

4-2. How to find the linkages between sustainability and business value

In the previous section, we introduced measurement which is clearly a significant step but managers should not stop there. If they do, they will only get a static sense of how sustainability is "working" Stakeholder centricity requires a deeper understanding of the connections between stakeholders' perceptions and behaviours.

Establishing linkages comes down to examining whether a change in one aspect of the framework (i.e., a variable) is related to changes in other variables. Sometimes, this sort of linkage is brought to light based on anecdotal evidence or the managers' experience. More often, though, the complex and highly psychological nature of stakeholder reactions to sustainability makes anecdotal evidence alone an insufficient basis for effective decision making. Managers should strive to base decisions on careful analysis that links variables in the framework and reveals the reasons behind stakeholder behaviours.

There are two essential steps in validly establishing these linkages, the first of which entails sound study design. This involves collecting data in ways that can shed light on the process. Primary data collected through surveys, laboratory experiments, and field experiments are typically necessary to shed light on the underlying psychological process, although sometimes secondary data collected at checkout counters etc., can also provide rudimentary evidence. As one illustration of study design, as mentioned in other chapters, we used a modified before/after experimental design to evaluate the impact of a million dollar donation by a company to a university. In this instance, we collected data through random surveys of stakeholders before and after the launch of the gift. The findings were clear that stakeholders who became aware of the company's gift had significantly greater intent to apply for jobs, purchase products, and invest in stock of the company than those who were unaware, whether that lack of awareness was due to being surveyed before the announcement, or simply because they had not been reached with the communications campaign. Thus, the experimental study design established that the program succeeded in driving a number of company favouring behaviours on the part of stakeholders.

In establishing linkages in our framework, the second necessary step is the use of appropriate statistical techniques. These techniques may be highly advanced or as basic as computing a correlation coefficient, but most techniques are designed to test whether increases (or decreases) in one or more variables result in, or are at least associated with, systematic changes in other variables. For example, in one set of study findings conducted with researcher Shuili Du, increases in Understanding for Stonyfield yogurt consumers were significantly related to corresponding increases in Unity (Du et al., 2007). Both Understanding and Unity were measured using 7-point scales, where 7 represented the most positive Understanding and the highest level of Unity. All other factors being equal (e.g., beliefs about product quality), an increase in Understanding of 1 point on the 1—7 scale was associated with an increase of .3 on our Unity scale. Thus, it became clear how much the company needs to improve Understanding in order to reach a targeted level of Unity.

When conducted correctly the above two steps can help managers:
- Establish the key drivers of desired behaviours
- dentify additional influencing factors

4-2-1. How to establish key drivers of business value

The key drivers of business value are the levers of the framework: Understanding, Usefulness, and Unity. So linking sustainability interpretations to business value is a matter of establishing correlations between these drivers and the desired outcomes. The most straightforward way to establish this is to measure both drivers and desired outcomes for a sample of individual stakeholders. For example, one would record a stakeholder's level of Understanding, Usefulness, and Unity with the company as well as that person's desired behaviour such as purchase intent, willingness to pay, or some other desired outcome. Relationships between drivers and outcomes can be analysed using standard statistical methods (e.g., multiple regression).

An experiment design may be essential to test sustainability concepts that have not yet been launched. For example, the researcher can have stakeholders view one of two or more descriptions of a program, varying the sustainability issue, sustainability implementation, or sustainability resources, or in some cases all three. Then the researcher can compare responses of stakeholders who saw one version to responses of those who viewed another version. For example, in an experiment that examined the business value of using organic materials in the manufacture of cotton t-shirts, researchers Trudel and Cotte (2009) presented potential consumers with varying levels of sustainable production (100% organic cotton, 50%, 20%, no ethical information given, unsustainable production). Consumers were willing to pay significantly less for a t-shirt from a company that used unsustainable production practices than one which used sustainable practices. Interestingly, the amount that consumers were willing to pay for t-shirts did not differ based on the degree to which organic materials were used, the willingness to pay for a 100% organic cotton t-shirt was negligibly higher than what they would be for one made from 25% organic cotton. This suggests that as long as some sustainability is detected by consumers, business value can be generated, but that using 100% organic cotton is not necessary to generate business value based solely on this metric (i.e., willingness to pay).

4-2-2. How to identify additional influencing factors

Stakeholder centric companies successfully adapt sustainability initiatives to the context in which they are conducted. There are two levels of contextual factors that managers need to measure, track, and incorporate in their sustainability decision-making, and then again later (as shown in the section below on calibrating), when they calibrate the effectiveness of sustainability activities. These two factors are the stakeholder context and the company context.

・Stakeholder factors.

Questionnaires are often employed to measure these factors. We have made these comparisons using both qualitative and quantitative data, and using a combination of statistics and careful study design. For example, in the study regarding a company's diversity program, the stakeholder's level of caring was calculated by listing a number of sustainability issues and asking the extent to which the respondent supported each issue. (Scale: 1=do not support at all; 7=strongly support.). The list included ten items such as equal opportunity employment practices, special employment support for gays and lesbians, special employment support for disabled people, and special educational opportunities for ethnic minorities. All these ratings were averaged into a single indicator of their caring for the sustainability issue of diversity.

In the qualitative case, we asked employees in various locations of a Fortune 500 company about the Usefulness of sustainability programs. The next step was coding the transcripts from each focus group and comparing responses for employees in various locations. This analysis revealed that Usefulness varied greatly depending on where the employee worked. Those at headquarters found sustainability to be Useful when it helped them become a better employee, by offering them an opportunity to learn new skills, for instance; but those located at an overseas office in a country where the local population was largely hostile to the company found programs most Useful when the program helped them deflect criticism from their social circles.

In a more quantitative case, we explored the beneficiaries of the previously discussed oral care program. In that study, those who participated in the sustainability program had enhanced levels of Understanding and were more likely to purchase the toothpaste brand in the future than non-participants. But going a step further, a set of differences across stakeholder characteristics became clear by comparing the responses of those participants who had lived in the US the shortest amount of time (low acculturation) to those who had lived in the US for the greatest amount of time (high acculturation). Due in part to the more recent generation's desire to fit in to American culture, the low acculturation participants showed enhanced Understanding and greater intent to reward the company through purchase than the highly acculturated group. Thus, although the sustainability program draws all participants closer to the company, the study revealed that the program proves to be especially effective for those who came to the US recently.

- **Company factors.**

Our research shows that business value accrues to companies in idiosyncratic ways; in other words, the same sustainability initiative may produce very different results at one company than another. Some of this fluctuation is due to the effects of industry, company size, or other fairly objective differences between companies. But in many cases, company context effects are more perceptual. For example, a stakeholder's reaction to sustainability will depend on how knowledgeable he or she is about the company and its competencies. One study examined the influence of the fit of the core business with sustainability initiatives (Becker-Olsen et al., 2006). That experiment randomly assigned respondents so that they evaluated a sustainability initiative that was either conducted at a company with a core business that was close to the sustainability initiative or not close at all. The researchers found that respondents were more sceptical of the company with the lower fit that that with higher fit.

4-3. Ways to calibrate CR and stakeholder relationship

Managing sustainability strategically requires a process whereby programs are calibrated regularly for constant improvement. Quantifying and finding linkages are important steps in sustainability management because they generate knowledge that can be leveraged so that stakeholder relationships are enhanced to the fullest extent possible. But the knowledge generated by these steps must be utilized properly to calibrate sustainability initiatives so that they fully improve understanding, satisfy stakeholder needs, and produce maximal business value.

There are three main applications of this knowledge—with existing CR programs, with future programs, and in other functional areas of the company that could benefit from improving stakeholder relationships. The latter two will spring from assessing the current programs and sharing that information with relevant others.

4-3-1. The dashboard: How to apply knowledge to improve sustainability initiatives

Aggregating all knowledge about each program on an ongoing basis is a necessary first step towards calibrating sustainability initiatives and stakeholder relationships. Leading companies create dashboard-like systems where managers can find all relevant information in a single source.

Like the well-known "Balanced Scorecard," (Kaplan and Norton, 1993) which focuses on sets of business performance measures ("Financial," "Customer," "Internal Business Processes," and "Learning and Growth"), a CR dashboard can be built upon key indicators for each step in our framework. This one has line items for Understanding, Usefulness, Unity, and two forms of business value (business and societal). The dashboard not only records current performance on these measures, but it also provides a target for the next period.

The dashboard can be used to improve programs—systematically and continually. Learning what is working and what is not can take time. But stakeholders' reactions to CR can evolve over time as well. Very often the fruits of a sustainability initiative do not become evident for weeks or even months after their introduction. For this reason, dashboards can track progress against tangible, agreed upon goals, so that every sustainability initiative is performing up to its fullest.

Gap analyses can be extremely useful when assessing sustainability strategy. When goals are not met, apparent by a gap between targets and actual results, managers can delve into underlying components of Understanding, Usefulness, and Unity for clues as to why the program is underperforming. In many cases, managers can re-calibrate the program based on this information, by improving communication to raise Understanding, or adding features to heighten Usefulness. If enhancing these levers is impractical or overly costly, jettisoning the program may be the best course of action.

4-3-2. Portfolio optimization of sustainability initiatives by leveraging knowledge

In creating sustainability strategy, quantification and finding linkages are important steps due to the fact that they generate knowledge that can be leveraged so that stakeholder relationships are enhanced to the fullest extent possible. Leading managers see sustainability as a means to reach broad business and societal goals through stakeholder relationship-building. For this reason, CR programs should not be viewed as standalone endeavours, but rather as a collection or portfolio of initiatives where knowledge is shared so that it may be leveraged by other CR management teams.

It may be apparent that this sort of sharing should go on within an organization; however, it is all too common for initiatives in marketing, operations, research and development, and corporate affairs to carry on without involving each other in our experience. For this reason, tools like the sustainability evaluation dashboard need to be made available to a wide variety of decision-makers. The British retailer Marks & Spencer is an example of a company that involves all departments and its employees in its sustainability strategy: It measures its performance against 180 social and environmental targets along the entire value chain and communicates the status of accomplishment in real time to all employees at its London headquarters via an electronic ticker (Hollender and Breen, 2010).

Knowledge sharing among sustainability managers is important for two reasons. First, our multi-step framework, from exposure to interpretation to response, is somewhat complex, making it unlikely that a manager familiar with only on sustainability initiative will have a realistic picture of stakeholder reactions to the company's overall sustainability strategy. Clearly, any sustainability team can develop more effective sustainability strategy if they have access to the latest knowledge gleaned from other initiatives. Upper-level managers need to ensure that as many sustainability managers as possible have access to up-to-date knowledge of how stakeholders react to sustainability so that future adjustments can be made by considering all the available information.

The second, less obvious, reason that sharing knowledge is important is that many stakeholders are exposed to numerous sustainability programs simultaneously. Since many stakeholders expect companies' actions to be coherent, lack of coordination between sustainability programs can lead to confusion, and ultimately suspicion of motives among a stakeholder base.

Therefore, one can perceive sustainability strategy as a portfolio of initiatives all of which have a role in providing value to a particular stakeholder. Leading companies manage these roles and generate synergies among sustainability programs by encouraging knowledge sharing.

5. Conclusion

It should be clear by now that sustainability management is a highly complex and context driven endeavour; mastering it is equally art and science. Stakeholder-centric companies seek input and involvement from stakeholders in the sustainability strategy (co-create), provide stakeholders with the right information about sustainability efforts (communicate), and continually track and adjust initiatives to maximize business value

1 The Stakeholder Route to Successful Sustainability Management   41

(calibrate). Sustainability strategy is thus better thought of as a process than a destination.

Scholars of customer centricity will tell you that customer centric companies not only understand what the customer values, but also how valuable the customer is to the company. These companies do more than merely try to be friendly to customers, they seek to serve the fundamental needs of those customers who will be most profitable in the long-term.

Using sustainability to become a stakeholder centric company serves this exact same purpose. Managers must identify which stakeholders are most important to the long-term health of the company and find ways to build long-term relationships with them. Our approach, leveraging the three U's of Understanding, Usefulness, and Unity is designed to do just that. It provides a window into the psychological process through which stakeholders interpret, evaluate, and act upon sustainability initiatives.

This article draws on the book Leveraging Corporate Responsibility: The Stakeholder Route to Maximizing Business and Social Value (2011), Cambridge University Press

<References>
Becker-Olsen, K. L., Cudmore, B. A. and Hill, R. P. (2006) 'The impact of perceived corporate social responsibility on consumer behavior', *Journal of Business Research,* Vol. 59, No. 1, pp. 46–53.
Bergami, M. and Bagozzi, R.P. (2010) 'Antecedents and Consequences of Organizational Identification and the Nomological Validity of the Bergami-Bagozzi-Scale', Working paper, Marketing Department, Jones Graduate School of Management, Rice University.
Bhattacharya, CB, Sen, S. and Korschun, D. (2008) 'Using Corporate Social Responsibility to Win the War for Talent', *Sloan Management Review,* Vol. 49, No. 2, pp. 37–44.
Bonini, S., Koller. T. M. and Mirvis Ph. (2009) 'Valuing Social Responsibility Programs', *McKinsey on Finance,* Vol. 32, pp. 11–18.
Cohen, C. and Prusak, L. (2001) *In Good Company: How Social Capital Makes Organizations Work,* Boston: Harvard Business School Press.
Cone (2007) 'Cause evolution survey', Available at: http://www.conecomm.com/2007-cause-evolution-and-environmental-survey, Accessed January 12, 2015.
Dawkins, J. (2004) 'Corporate Responsibility: The Communication Challenge,' *Journal of Communication Challenge,* Vol. 9, No. 2, pp. 108–119.

Dwyer, F. R., Schurr, P. H. and Oh, S. (1987) 'Developing Buyer-Seller Relationships', *Journal of Marketing,* Vol. 51, No. 2, pp. 11-27.

Du, S., Bhattacharya, CB. and Sen, S. (2007) 'Reaping Relationship Rewards from Corporate Social Responsibility: The Role of Competitive Positioning', *International Journal of Research in Marketing,* Vol. 24, No. 3, pp. 224-241.

Fombrun, C. and Shanley, M. (1990) 'What's in a Name? Reputation Building and Corporate Strategy,' *Academy of Management Journal,* Vol. 33, No. 2, pp. 233-258.

Greening, D.W. and Turban, D.B. (2000) 'Corporate Social Performance as a Competitive Advantage in Attracting a Quality Workforce,' *Business & Society,* Vol.39, No.3, pp. 254-280.

Haley, E. (1996) 'Exploring the Construct of Organization as Source: Consumers' Understanding of Organizational Sponsorship of Advocacy Advertising,' *Journal of Advertising,* Vol.25, No.2, pp. 19-36.

Hill, R. P., Ainscough, T., Shank, T. and Manullang, D. (2007) 'Corporate Social Responsibility and Socially Responsible Investing: A Global Perspective', *Journal of Business Ethics,* Vol. 70, No. 2, pp. 165-174.

Hollender J. and Breen, B. (2010) *The Responsibility Revolution,* San Francisco: Wiley and see also, http://plana.marksandspencer.com/about, accessed January 12, 2015.

Kaplan, R. S. and Norton, D. P. (1993) 'Putting the Balanced Scorecard to Work', *Harvard Business Review,* Vol. 71, No. 5, pp. 134-147.

Korschun, D. and Du, S. (2013) 'How Virtual Corporate Social Responsibility Dialogs Generate Value: A Framework and Propositions', *Journal of Business Research,* Vol. 66, No. 9, pp. 1494-1504.

KPMG (2008) 'KPMG International Survey of Corporate Responsibility Reporting', Available at: http://www.kpmg.com/EU/en/Documents/KPMG_International_survey_Corporate_responsibility_Survey_Reporting_2008.pdf, Accessed January 12, 2015.

Lawrence, A. T. (2010) 'Managing Disputes with Nonmarket Stakeholders: Wage a Fight, Wait, Withdraw, or Work It Out?', *California Management Review,* Vol. 53 No. 1, pp. 90-113.

Lichtenstein, D. R., Drumwright, M. E. and Braig, B. M. (2004) 'The Effect of Corporate Social Responsibility on Customer Donations to Corporate-Supported Nonprofits', *Journal of Marketing,* Vol. 68, No. 4, pp. 16-32.

Menon, S. and Kahn, B. E. (2003) 'Corporate Sponsorships of Philanthropic Activities: When Do They Impact Perception of Sponsor Brand?', *Journal of Consumer Psychology,* Vol. 13, No. 3, pp. 316-327.

Morgan, R. M. and Hunt, S. D. (1994) 'The Commitment-Trust Theory of Relationship Marketing', *Journal of Marketing,* Vol. 58, No. 3, pp. 20-38.

Petty, R. E., Cacioppo, J. T. and Goldman, R. (1981) 'Personal Involvement as a Determinant of Argument-Based Persuasion', *Journal of Personality and Social Psychology,* Vol. 41, No. 5, pp. 847-

855.

Rosen, S., Irmak, C. and Jayachandran, S. (2009) 'Value Co-Creation in Cause-Related Marketing: How Letting Consumers Choose the Cause Enhances Consumer Support', Working Paper, Moore School of Business.

Sen, S. and Bhattacharya, CB. (2001) 'Does Doing Good Always Lead to Doing Better? Consumer Reactions to Corporate Social Responsibility', *Journal of Marketing Research,* Vol. 38, No. 2, pp. 43–62.

Sen, S., Bhattacharya CB. and Korschun, D. (2006) 'The Role of Corporate Social Responsibility in Strengthening Multiple Stakeholder Relationships: A Field Experiment', *Journal of the Academy of Marketing Science,* Vol. 34, No. 2, pp. 158–166.

——, Du, S. and Bhattacharya, CB. (2009) 'Building Relationships through Corporate Social Responsibility', in MacInnis, D. J., Park C. W. and Priester, J. R. (Eds.) *Handbook of Brand Relationships,* M. E. Sharpe.

Simmons C. J. and Becker-Olsen, K. L. (2006) 'Achieving Marketing Objectives through Social Sponsorships', *Journal of Marketing,* Vol. 70, No. 4, pp. 154–169.

Target (2008) 2007 Corporate Responsibility Report, Available at:

https://corporate.target.com/_ui/pdf/2007_full_report.pdf, Accessed January 12, 2015.

The Harris Poll Reputation Quotient (2012) Available at:

http://www.harrisinteractive.com/Products/ReputationQuotient.aspx, Accessed January 12, 2015.

Trudel, R. and Cotte, J. (2009) 'Does It Pay To Be Good?', *MIT Sloan Management Review,* Vol. 50, No. 2, pp. 61–68.

Yoon, Y., Gürhan-Canli, Z. and Schwarz, N. (2006) 'The effect of corporate social responsibility (CSR) activities on companies with bad reputations', *Journal of Consumer Psychology,* Vol. 16, Issue 4, pp. 377–390.

Whole Foods Market (2012) Available at: http://www.wholefoodsmarket.com/value, Accessed January 12, 2015.

〈翻訳〉1 サステナビリティ経営を成功させる ステイクホルダー・ルート

CB バタチャリア

Professor, European School of Management and Technology in Berlin, Germany

(JFBS 事務局訳)

【要旨】
本稿では，ステイクホルダーに重点をおき3つのポイントについて論じる。まず，企業が自分たちとともにサステナビリティを実現していくようステイクホルダーに働きかける方法を示す。それはステイクホルダーの注意を企業のサステナビリティ活動に引きつけ，より深く「理解」してもらうこと(Understanding), 自分たちのプログラムがすぐれて「有用」であることを保証すること(Usefulness), 企業とステイクホルダーが同じ価値観をもっているというステイクホルダーの感覚，つまり「統合」を強めること(Unity), という3つのUから成るプロセスである。次に，コミュニケーションが果たす重要な役割を指摘する。ここでいうコミュニケーションとは，ステイクホルダーの企業に対する親近感を増すような方法で情報発信を行うことである。最後に，企業が長期的にサステナビリティ活動を調整・実施していくことができるように，重要な要因を定量化し要因間の関係を明らかにする方法について手引きを示す。

1. はじめに

サステナビリティ経営の成功によって，企業業績は大きく向上する。たとえばCO_2放出量の削減によってエネルギーコストを節減したり，生産活動から生じるゴミの削減およびリサイクルによってサプライヤーからの原材料購入量を低減したりするように，サステナビリティ活動そのものが業績向上に結びつくことがある。こうした取り組みは，サステナビリティ活動が企業価値を生む

直接的ルートと考えられている。この直接的ルートは多くの場合見えやすく，測定が比較的容易である。

しかしここで示したいのは，企業価値をもたらすもうひとつの補完的なルート「ステイクホルダー・ルート」である。ステイクホルダーが取り組みについて知り，反応することによって企業価値が生まれるルートであり，（直接的ルートとは対照的に）見えづらく測定が難しい場合が多い。Bhattacharya, Sen and Korschun (2011) *Leveraging Corporate Responsibility: The Stakeholder Route to Maximizing Business and Social Value* において詳細な検討を行っているが，本稿では企業業績につながるこのステイクホルダー・ルートに焦点を当て，概説していくことにしよう。

自社の可能性を引き出しステイクホルダーと強固な関係を構築する企業は，短期的にも長期的にも便益を得ることができる。経営者およびサステナビリティの専門家を対象とした調査によれば，ステイクホルダー・ルートは，評判の確立，従業員のやる気の向上，競争優位の強化など，多様な方法で企業価値を生み出す可能性がある（Bonini et al., 2009）。企業の優れたサステナビリティ活動によって消費者のロイヤリティが培われていき，彼らが企業の社会的方針を支持し少々割増しの価格でも厭わず購入するブランド・アンバサダーや支援者になることもある（Sen and Bhattacharya, 2001）。サステナビリティはまた，有能な従業員（Greening and Turban, 2000）や社会的に責任ある投資家（Sen, Bhattacharya and Korschun, 2006; Hill et al., 2007）を引きつけ，動機づけ，囲い込む上で競争優位を生む可能性がある。

ステイクホルダー・ルートが生み出す便益は，ステイクホルダーのニーズや要求を理解し，それに応えていく能力を企業がどれだけ有するかによって異なるだろう。このような考え方を「ステイクホルダー中心主義」と呼ぶことにする。消費者，従業員，投資家，地域コミュニティなどの声に敏感な企業は，サステナビリティへの取り組みによって多くの便益を得ることになる。対照的に，ステイクホルダーの潜在的な動きを無視する企業は，ステイクホルダーとの関係に傷がついてしまい相当なダメージを被ることになるだろう。本稿では，ステイクホルダー中心主義の企業がステイクホルダーと強固な関係を構築するた

めに，3つの重要なサステナビリティ活用方法について検討を行う。第一に，企業とともにサステナビリティを実現していくようにステイクホルダーに働きかける方法を示す。第二に，企業に対するステイクホルダーの親近感を増すようなコミュニケーションの役割を指摘する。第三に，企業が長期的にサステナビリティ活動を調整・実施していくことができるよう，重要な要因を定量化し要因間の関係を明らかにする方法を示す。

2. サステナビリティ戦略の共創

われわれは，サステナビリティ戦略への共創アプローチを提唱している。実際，戦略策定にステイクホルダーによるエンゲージメントを求めない企業がステイクホルダー中心主義であるとは考えられない。CSRプログラムを立案する際，ステイクホルダーを適切に巻き込むことによって焦点を絞ることができ，より大きな成功を収めることができる。共創によってステイクホルダー自身が解決策を講じるため，ステイクホルダーがもっている期待と企業の対応とのギャップが低減するからである。企業はまずサステナビリティをうまく共創するための要因を理解し，ステイクホルダーが十分にサステナビリティを共創できるプロセスを構築しなければならない。

2-1. 成功する共創戦略のための要因

われわれは，効果的なサステナビリティ戦略を共創する企業が，サステナビリティ活動の中心的存在として従業員および消費者に働きかけを行っていることを明らかにしている。たとえば，チャリティのために従業員から寄せられた寄付金に企業が同額を上乗せするマッチングギフトプログラムがある。これは従業員が支援対象のコーズ（例：環境団体やコミュニティのサステナビリティ）や寄付金額を個々に決めることができるため，従業員に評判がよい。しかしながら，サステナビリティ活動の実施を成功させる主要な要因が明らかであるとは言えない。

われわれは，これまでの研究および経営者への聞き取り調査によって，CSR

プログラムの実行戦略を成功させる3つの要因を特定した。

1) ステイクホルダーが能動的な行為者であるか，受動的な行為者なのか，
2) 社内の他の従業員やプログラムの関係者との関係が水平的か，垂直的か，
3) 経営者はプログラムの評価プロセスを公式化しているか，非公式か。

　経営者にとって，これらの要因はサステナビリティ戦略を実行する際の手引きとなるものである。企業がサステナビリティへの取り組みをステイクホルダー中心主義的に計画していくために，われわれの研究が指摘しているのは，経営者はステイクホルダー間の非公式な関係性を高めていく公式なプロセスを導入するべきであるということだ。このプロセスは，ステイクホルダーにとってはやりがいを感じる経験になるだけではなく，経営者にとってはステイクホルダーのニーズや選好をよく知るための市場調査にもなる。こうしてステイクホルダーの役割や任務を定義し，それを企業経営に組み込み，ステイクホルダーからの評価を獲得するのである。組織のモニタリング，評価，報酬のシステムをプロセスに含めることが重要である。

2-2. よりよい関係づくり

　ステイクホルダー中心主義および共創アプローチは，企業とステイクホルダー間の対話を促し，そこから得られたものを可能な限り早い段階からサステナビリティ構築プロセスに組み込んでいく。最も重要なのは，すべてのステイクホルダーの便益を最大化するサステナビリティ活動を共創することである。そのためには，ステイクホルダーを巻き込み，彼らがサステナビリティ活動から得る便益を増やす方法を検討することが必要である。このようなプロセスによって，ステイクホルダーにとっての付加価値をともに創りだしていくだけでなく，最終的に関連するすべてのプログラムや課題が考慮されることになる。

　経営者はこうした検討結果をどのように実行に移しているのだろうか。サステナビリティ構築をうまく進めるためには，明瞭化，創出，抽出，選定の4つ

の段階がある。明瞭化と抽出の段階では，経営者から提供される情報に依存するが，創出と選定の段階では，重要なステイクホルダーと積極的にかかわってサステナビリティに関する意思決定を行う共創作業が行われる。

2-2-1. 明瞭化

　サステナビリティ戦略を構築していく関係者にとって極めて重要なタスクは，どのようなサステナビリティ価値をつくり出したいのか決め，サステナビリティ活動の目標を公式に明瞭化することである。この段階を踏んではじめて，巻き込む対象のステイクホルダー・グループが決まる。最終目標が明確でなければ，サステナビリティ価値を実現する手段を決めることは非常に難しくなる。たとえば，従業員を引きつけ囲い込むことが目標である場合と比べて，企業評判やブランド価値を高めることが目標である場合は多様なステイクホルダーを対象にする必要があるなど，目標によってサステナビリティ戦略は相当異なる。

　サステナビリティ戦略の構築においては，明確な目標を定めること，自ら積極的に取り組む姿勢をステイクホルダーから引き出すような方法でその目標を明らかにする必要がある。経営者は，必要な資源を獲得するためにサステナビリティ活動の立上げ支援を積極的に行うとともに，確実に取り組みを行い結果を出すよう長期的にサポートし続ける必要がある。従業員とともにCSRの目標を定め，積極的に取り組む姿勢を社内から幅広く引き出すには，知見を得ながら従業員同士で共通の理解を形づくっていくことが大切である。

2-2-2. 創出

　対象となるステイクホルダーを特定し目標を設定した後は，目標に沿ったサステナビリティのプログラムの創出に向けて，企業はステイクホルダーとかかわり，交流していくことが重要である。個人的なニーズ（見えやすいニーズであれ，見えにくい心理的ニーズであれ）を掘り下げていく質的研究を通して，個々のステイクホルダーとかかわっていくことができる。

　社会的に責任ある企業とかかわっているステイクホルダーには，重要な心理的便益として自尊心と一貫性の感覚がもたらされるということが，われわれの研究で明らかになっている。サステナビリティは消費者自身の価値を高める。つまり消費者は，責任ある企業からモノを買うという行為によって満足感をも

ち，またそのモノを提供する企業が自分と同じ価値観をもっていると認識する。消費者は，それぞれの生活の局面（例：仕事，余暇，ひとりの時間）においても自己の一貫性を感じることになる。

　ステイクホルダー中心主義の本質であるステイクホルダーのニーズを理解することによって，経営者は共創に最も適した要因をつかむことができるのである。

2-2-3. 抽出

　ステイクホルダーとの協働によって社会的責任プログラムを創出した後，経営者は企業のコアコンピテンシーと合致した実現可能なものとなるように，その取り組みに優先順位をつける，あるいは重要なものを「抽出」する必要がある。それが正しく実行できなければ，最良のサステナビリティ活動ですら無駄な努力になってしまう。

　この段階では，マテリアリティ・マトリックスを作るとよい。マトリックスには，「ステイクホルダーにとっての重要性や魅力」の次元と，サステナビリティへの取り組みが事業の成功に及ぼす影響という意味での「企業にとっての重要性」という次元を示す。創出段階で共創された選択肢群をふるいにかけ，企業がうまく実行できるように絞っていくのである。

2-2-4. 選定

　明瞭化，創出，抽出に続き，実行に向けた選定の段階がある。経営者は，最終的な選定段階において最善の取り組みを選び，各取り組みのターゲットとなるステイクホルダー・セグメントを確定することになる。ステイクホルダー中心主義の企業は，ステイクホルダーがどのように理解しどのような態度をとるかということを第一（唯一ではない）にして選定を行う。様々な取り組みが生みだす価値を定量化するために，ステイクホルダーに関する追加的情報としてステイクホルダーサーベイが行われる。このデータは，経営者がどの取り組みを実行するか最終決定を行うために用いられる。

　この選定段階では，サステナビリティ・プログラムに関する社内の責任者が，プログラム実施パートナーとなるNPOとの連携を開始する。企業のメンバーとパートナーNPOから成るチームは，

・コーズやコーズの実行方法（例：マーケティング，ボランティアなど），
・パートナー候補となる NPO，
・提案されたステイクホルダーエンゲージメントのタイプとレベル，

などでサステナビリティ活動を具体化していく。企業と NPO は長期にわたるパートナーシップを構築するため早いうちから共創の土台を築く必要がある。そうした働きかけが NPO や他のステイクホルダーとの共創を生むことになる。

3. サステナビリティ戦略のコミュニケーション

　企業がサステナビリティへの取り組みをステイクホルダーと共創することによって，いかにステイクホルダー中心主義になっていくことができるかを前節で示した。サステナビリティ活動の共創は，ステイクホルダーにとっての便益を増やすだけでなく，彼らがサステナビリティ活動を深く理解することにより企業に便益をもたらすことになる。ステイクホルダー中心主義とは，ステイクホルダーが企業のサステナビリティの取り組みを知ることで，企業に非金銭的な便益をもたらすことを意味している。そこで次にステイクホルダーと企業が強固で継続的な関係を構築できるよう，サステナビリティへの取り組みに関するコミュニケーション戦略を検討していこう。戦略が成功するには，コミュニケーションの理由，対象，場所をはっきりさせ，企業やステイクホルダーの置かれた特別な状況を把握することが求められる。

3-1. コミュニケーションにおける重要な課題

　コミュニケーションは，ステイクホルダーにとってサステナビリティ活動に関する理解を深める主要な手段であり，サステナビリティ戦略の重要な要素である。また，ステイクホルダーがサステナビリティ活動を理解し，その企業とのつながりを深めていくかどうかを決める出発点でもある。コミュニケーションの役割は，すべてのサステナビリティ・プログラムに直接かかわれない多くのステイクホルダーのために，企業にとって最も価値を生み出す活動の情報を発信することである。

効果的コミュニケーションを模索する経営者は，ここで2つの課題に直面する。個々の取り組みに関するステイクホルダーの認識の低さと，（状況は変わりつつあるとはいえ）サステナビリティ・プログラム開始当初に多くのステイクホルダーがもつ懐疑心である。

・ステイクホルダーの認識

いうまでもなく，企業のサステナビリティ活動に対するステイクホルダーの認識が低い限り，サステナビリティがステイクホルダー・ルートを通じて企業価値を生み出すことは難しい。われわれの研究によれば，企業の社会・環境への取り組みに対する認識は極めて低い。ある調査では，消費者のうちわずか16％ほどが認識しているにすぎず，別の調査（雇用者を対象とした全国的調査）では少し高い水準とはいえ，約半数の従業員のみが企業の社会・環境への取り組みを認識しているにすぎなかった。また，市場調査会社 Harris Interactive が企業の評判に関する年次調査の中で行った20指標に基づく調査でも，CSRについては「わからない」という回答が最も多かった。企業は社会的・環境的責任を果たしているかどうかという問いに対しても，大半が「分からない」と回答した（The Harris Poll Reputation Quotient, 2012）。このように企業が社会的責任を果たすためのプログラムに取り組み，そこから戦略的に成果を得ようとしても，ステイクホルダーの認識の低さが障害となっている（Sen et al., 2006; Du et al., 2007; Bhattacharya et al., 2008）。

・ステイクホルダーの懐疑心

自らがかかわるサステナビリティ戦略を積極的に知ろうとしているステイクホルダーは多いが，この熱心さを自社に対する関心だけからくるものだと誤解してはならない。企業不祥事，グリーンウォッシュ，サステナビリティへの取り組みの誇張（Sen et al., 2009）などに関するメディアの報道に晒されうんざりしている人々は多い。経営者にとってステイクホルダーがもつ懐疑心を取り除くことは，労力を要する課題である。

われわれの研究を含め先行研究が示すとおり，ステイクホルダーがサステナビリティ活動について知ろうとする時は，何が企業を動機づけているのか―本当に社会的・環境的課題への関心からかそれとも利益追求のためか―というこ

とを判断しようとする。Yoon et al. (2006) は,コミュニケーションを通じて企業の動機を説明する必要があると指摘している。サステナビリティ活動および活動から得られた利益についての情報発信を控える企業もあるが,本稿ではそうしたアプローチを提唱している訳ではない。本当に社会的・環境的課題への関心に基づいてサステナビリティ活動が行われているのであれば,利益追求の動機もステイクホルダーに受け入れられることが,Sen et al. (2006) のサステナビリティに対する関心の研究により明らかになっている。実際,ステイクホルダーの中には環境とビジネスの両方にメリットのあるサステナビリティ活動を革新的であると賞賛し,企業に便益をもたらしている。

3-2. メッセージの内容

　サステナビリティ・コミュニケーションは,社会的なコーズに対する企業の取り組みに焦点を当てている。企業が強調する要因には,コーズへのコミットメント,そのコミットメントがうみだしたインパクト,コーズと企業の本業の一致などがある。

・サステナビリティへのコミットメント

　サステナビリティ・コミュニケーションにおける重要な最初のステップは,企業がコーズにコミットしていることを示すことである。ステイクホルダーが企業のコミットメントを評価する指標には一般的に次の3つがある。

- ・インプットの量（例：寄付金額,現物供与,マーケティング専門知識や人的資本）
- ・インプットの一貫性（Dwyer et al., 1987）
- ・コミットメントの継続期間

　この中で経営者が最も過小評価するのは,コミットメントの継続期間である。長期間にわたるコミットメントは容易でないため,信頼される。これを裏付ける研究によれば,短期間のキャンペーンと比べてコミットメントが長期間になればなるほど,ステイクホルダーは企業が本当の関心をもって社会や地域の状況を改善するために活動しているとみなす傾向があった。

・サステナビリティのインパクト

　ステイクホルダーは企業がサステナビリティの課題に深くコミットすることを期待し，またそのサステナビリティ活動が社会の状況を実際に改善していることを示す証拠を求めている。経営者はインプットに見合うサステナビリティ成果を挙げることにより，ステイクホルダーの期待に応えることができる。企業のサステナビリティ活動による社会的インパクトと，消費者の企業活動に対する関心との間には，正の相関があることが先行研究により明らかにされている。

・サステナビリティとの一致

　最後に，コミュニケーションの重要な要因としてサステナビリティの課題と本業との一致が挙げられる。ステイクホルダーは本業に沿った課題を支援することを企業に期待している（Cone, 2007; Haley, 1996）。製品の特徴（例：ハーブ製品を販売するブランドによる熱帯雨林の保護の支援），特定のターゲット層に対する親近感（例：エイボンと乳がんキャンペーン），特定の社会領域で過去に実施したプロジェクトによってつくられた企業イメージ（例：ベン＆ジェリーとボディショップの環境保護活動）などは，サステナビリティの課題と本業の一致を図る事例である（Menon and Kahn, 2003）。

　CSRとの一致もまた，ステイクホルダーの企業への関心に影響をおよぼす重要な要因である（Menon and Kahn, 2003; Simmons and Becker-Olsen, 2006）。サステナビリティ活動と本業の間に論理的な関係が見出せないとき，ステイクホルダーは企業の本当の動機は何かと疑問をもつ。サステナビリティの課題と本業が一致すれば，ステイクホルダーの企業に対する理解は深まり，コーズの背景にある企業の本当の関心に対する理解も深まる。当初その一致の度合いが低い場合でも，ステイクホルダーに受け入れられるように変えていくことはできる。

3-3. コミュニケーションのチャネル

　サステナビリティ活動に関する情報や記録は，多様なチャネルを通じて広まる。一般的には，企業自身がコントロール可能な企業のチャネルと企業にはコ

ントロール不可能な外部のチャネルがあり，それぞれを適宜組み合わせていくことにより取り組みを伝えていく。

・企業のチャネル

企業のチャネルでは，サステナビリティ活動のとりまとめ・報告のほとんどを企業自身が担う。対象とするステイクホルダーによって様々なチャネルが用いられる。企業のチャネルとしてよく用いられるものとしては次のようなものがある：

・サステナビリティレポート
・ウエブサイト
・E-mail による発表
・広告
・プレスリリース
・商品パッケージ

KPMG の調査（2008 年）によると，世界の大企業の約 80％が CSR 報告書を発行しており，2005 年の約 50％を大きく上回っている。ダイエットコーラは心臓病に対する女性の関心を高めるテレビ広告を打っており，このコミットメントを伝えるためのウエブサイトも開設した。またストーニーフィールドは，健康と環境に対する多様な取り組みをステイクホルダーに伝えるメッセージをカップヨーグルトのふたに印字している。

企業がもつチャネルは，企業戦略およびサステナビリティ戦略と矛盾なく情報発信できるため，企業にとっては魅力的なチャネルである。ステイクホルダーに届けるメッセージを管理し，企業のコミットメントとそれがうむインパクト，そしてサステナビリティとの一致を最適な形で伝えることができるため，ステイクホルダーがもつ特定のニーズに対してもより良い形で取り組んでいくことができるだろう。ただこのチャネルには，他のコミュニケーション方法と比べて信頼を得にくいというマイナス面がある。先行研究によると，消費者は企業から独立した情報源から入手する情報に肯定的であり，懐疑的な反応が弱いことが示されている（Yoon et al., 2006; Simmons and Becker-Olsen, 2006）。つまり，ステイクホルダーは第三者から発信される情報より企業から発信される

情報に懐疑心をもちがちであると言える。

・外部のチャネル

　企業がもつチャネルと対照的なのが，現在大きく増えつつある外部のチャネルである。メディア，消費者，モニターグループ，消費者のブログなどは，広範囲にわたるステイクホルダーとコミュニケーションを図ることができる外部チャネルのほんの一例である。これらは企業の直接コントロール下にはないため，発信する情報が企業のサステナビリティ戦略と一致しないかもしれない。しかしながらその独立性ゆえ，ステイクホルダーは企業がもつチャネルよりも外部のチャネルから受け取る情報を信頼する傾向がある。企業は，自社のサステナビリティ戦略と外部のチャネルから発信される情報との間の整合性がとれるよう，ステイクホルダーとの対話を怠ってはならない。

　ソーシャルメディアの絶え間ない発展により，誰もが企業に関する情報の発信者になることができる。消費者はフェイスブックに意見を投稿するほか，企業からの情報を他者へリツイートすることも可能である。ストーニーフィールドファームやベン＆ジェリーのような企業は，バーチャルな世界で社会的責任プログラムを称賛する消費者アンバサダーから便益を得てきた。また従業員は，社会的ネットワークを通して他のステイクホルダーとの幅広いアクセス機会をもっており，信頼に足りる情報源とみなされる (Dawkins, 2004)。先進的企業はサステナビリティ活動へのステイクホルダーの関心を集めるとともに彼らの期待に応えるため，ステイクホルダーとのバーチャルな対話に取り組んでいる (Korschun and Du, 2013)。

3-4. コミュニケーションの効果とそれを促すもの

　他のコミュニケーションと同様，サステナビリティ・コミュニケーションは企業それぞれに立案する必要がある。独自のコミュニケーションに工夫を凝らしている先進的企業が存在している。経営者が企業とステイクホルダーそれぞれの要因について考慮するに当って，次の2点が重要である。

・企業特有の要因

　サステナビリティ・コミュニケーションにより有利になる企業と不利になる

企業がある。ステイクホルダーは既にある情報に基づいて判断するため，評判の良い企業はステイクホルダーの反応という点において相対的に有利である(Fombrun and Shanley, 1990)。つまり，評判の良い企業ほどステイクホルダーの信頼性は高いため，サステナビリティ・コミュニケーションがもたらす肯定的効果はより高まる。対照的に，評判のよくない企業はサステナビリティ・コミュニケーションの効果が出ないばかりか，場合によっては逆効果になることもある。

例えばティンバーランドやベン＆ジェリーのような長期にわたってサステナビリティ・リーダーという評判をもつ企業は，サステナビリティ・コミュニケーションの効果は著しく高いと言える。

アメリカのスーパーマーケット業界の場合，ホールフーズマーケットは「わたしたちのコミュニティと環境を大切に」というコア・バリューを採択していることにより，CSR に関する地位を確立している。さらにこのコア・バリューは，オーガニック調達およびサステナブル調達を行う環境配慮型小売業として同社のすべての局面に浸透している。同社では年間利益の少なくとも5％を様々なコーズに充当し，従業員に対して就業時間内にコミュニティ活動に従事することを奨励している。

・ステイクホルダー特有の要因

企業側の要因に合わせてサステナビリティ・コミュニケーションを行うように，それぞれのステイクホルダーがもつ特有の性質や選好も考慮に入れなければならない。たとえば，NGO は企業 CSR プログラムがもたらす社会的インパクトについて証拠を強く求める傾向がある。対照的に，社会的に責任ある投資家はサステナビリティ活動がもたらすビジネスインパクト―顧客価値，従業員の定着率，コーポレート・ガバナンス，リスクマネジメントといった重要なビジネス指標と CSR プログラムがどのように関連しているか―についても社会的インパクトと同程度の関心をもっている。また専門的な関係者は，GRI や AA1000 などなじみのある情報開示基準に沿うことを求めるであろう。

これだけ幅広いステイクホルダーの中で，それぞれもっとも関心のあるコーズについての選好は異なる。人々は自分に関係ある情報には大いに関心をもっ

ていることを踏まえると，サステナビリティ・コミュニケーションが最も効果的なのはステイクホルダーが最も関心をもつコーズに取り組む時である（Petty et al., 1981）。このようなコミュニケーションは，メディアによる雑情報に左右されず，ステイクホルダーからの支持を取り付けるのに最も効果的な方法である。たとえばGAP社は，売上1ドルあたり5%分を6つのNPOから顧客自身が選んだ1つの団体に寄付するというコーズ・プロモーションを最近行っている。6つのNPOは，GAP社が幅広い社会的課題の中から代表的なものを選定した。選定されたNPOは，国内問題を代表する団体として「ティーチ・フォー・アメリカ」（教育問題）や「フィーディング・アメリカ」（国内の子どもの飢餓），グローバルな問題を代表する団体として「世界自然保護基金」（環境問題）や「グローバル・ファンド」（エイズ・結核・マラリアなどの疾患）などである。寄付先となる社会的課題とNPOをステイクホルダーに選んでもらうことによって同社は消費者から高い支持を取り付け，CSRコミュニケーションの効果を高めることができた。

4. サステナビリティ戦略の調整方法

　企業とのかかわり方を決めるに当って，つい最近までほとんどの消費者，従業員，投資家はサステナビリティを考慮に入れることはなかったが，今日では状況が一変している。ステイクホルダー中心主義の企業は，ステイクホルダーのニーズや選好に合わせるため，ステイクホルダーとともにサステナビリティ戦略を調整し続けることが求められる。

　企業はどのようにステイクホルダーの情報を把握するのか。サステナビリティ戦略を調整しながらサステナビリティ経営を実行可能にするためのポイントを考えていこう。次節では，社内で共有できるように価値を定量化しその結果を実行可能な知見にしていく手法について考えていく。その調整プロセスは次の3段階から成る。

　　・ステイクホルダーの認識と企業価値を定量化する
　　・サステナビリティと企業価値の関係を見い出す

・企業価値を最大化する機会を特定する

4-1. ステイクホルダーの認識と企業価値の定量化

有名な格言「測定できないものは管理できない」は，他のビジネス領域と同様サステナビリティ領域にもあてはまる。サステナビリティのために行った投資から得られる利益を最大化したいのであれば，経営者はステイクホルダーの反応（行動反応と心理的解釈）を測定する必要がある。

本研究では，CSRに対するステイクホルダーの反応を定量化するにあたり，フィールドおよび実験室で行われるサーベイと実験の両方を含め定評ある手法や技法を多く用いる。対象となるステイクホルダーを従業員，消費者，BtoB顧客などに細かくセグメント化しているが，企業価値に影響を及ぼす可能性があるステイクホルダーとの関係について，あらゆる視点から測定することを共通テーマとしている。このアプローチは次の2つの問いに答えるものである。

・企業のサステナビリティ活動をステイクホルダーはどのように解釈しているのか？

・企業がステイクホルダー・ルートによって求める成果とはどのようなものか？

4-1-1. サステナビリティに対するステイクホルダー反応の現状

サステナビリティ・エンゲージメントを最適化するには，ステイクホルダーの心理作用を正しく理解するために，彼らの心を理解する信頼性ある方法が必要である。ステイクホルダーの反応は，3つのコア・コンセプト―サステナビリティへの取り組みに対する理解，その取り組みが自分にとって有用であるとみなす度合い，企業に対して感じる統合―から成る心理プロセスを経て生まれる。強固で持続的なステイクホルダーとの関係は，これら3つのコンセプトの上に築かれるものであり，これらがうまく調和することで，企業業績の著しい向上につながると言える。

・理解（Understanding）

理解とは，簡潔にいえば，企業のサステナビリティについてステイクホルダーが抱いている認識のまとまりである。ステイクホルダーが企業のサステナ

ビリティについて知り，「この企業がよりサステナブルになるには何が動機となるのだろうか」，「サステナビリティへの取り組みは，企業が表明しているように社会に役立っているのだろうか」といった問いに答えることによって理解が深まる。

こうした問いが，企業がもつ関心やサステナビリティのインパクトから影響を受けて生じることは明らかであろう。そこでわれわれは，企業に対して本当の関心と利益追求のモチベーションの両方を測定するよう助言している。ステイクホルダーは，CSRを理解するにあたりこの2つを比較する傾向がある。これらが両立していれば革新的アプローチであると評価され企業に便益をもたらすため，両者の相互関係が明らかになるよう測定を行うべきである。理解を深めるもう一つの要因は，ステイクホルダーがサステナビリティ活動のインパクトをどのように知覚するかということであり，プログラムのレベルと企業レベルの2つのレベルで測定される。ステイクホルダーに「そのプログラムは効果があるか」，「どのような効果があるか」と質問することによって，特定のプログラムのインパクトが測定される。次に，サステナビリティ・プログラムのインパクトに関するステイクホルダーの知覚が同程度の企業同士を比較することによりその基準値が設定され，企業のCSRプログラムの全体的な有効性が測定される。

・**有用性**（Usefulness）

ステイクホルダーは，サステナビリティの有用性を見極めるよう絶えず経営者に対し働きかけることにより，さまざまな便益を得る。有用性には機能的便益と心理的便益があり，経営者は両方をみる必要がある。機能的便益は非常に具体的であり，サステナビリティ活動実施にともなう直接的な結果として現れる。たとえば，口腔ケアプログラムに関するわれわれの調査では，両親は子供たちがきれいで健康な歯を手に入れることになったという。他の例としては，電球型の蛍光灯の購入後消費者はエネルギー消費量を抑えることになったというものがある。

心理的便益は，自尊心や一貫性の感覚を保ちたいというステイクホルダーのアイデンティティ上のニーズから生まれる。こうしたニーズは，企業との関係

がいかなるものであるかにかかわらず，あらゆるステイクホルダーに共通している。しかしながら，ステイクホルダーは置かれた状況や自分の役割に沿って自分の感情を表すため，多様な心情をもつものとして測定を行う必要がある。たとえば，家庭生活では自尊心が高くても仕事においては非常に低くなる可能性がある。そのような状況では，職場での自尊心と家庭での自尊心のいずれか（あるいは両方）に関する質問をすることが望ましい。

・統合（Unity）

　統合とは，ステイクホルダー中心主義の中核にある，ステイクホルダーと企業の関係を如実に反映する総括的概念である。「企業は自分と同じ価値を共有しているか」，「将来的にも同じ価値観をもつといえるほど信頼できるのかどうか」とステイクホルダーが考えることである。顧客満足度のようなマーケティングの代表的な評価基準は，統合を考えるための視座を提供する可能性はあるが，企業の特性や「想い」に関するステイクホルダーの評価を可視化できないため統合を理解するには不十分である。

　そこで本研究では統合を次の2つの方法で測定している。1つ目は対象組織との一体感の度合い，すなわちステイクホルダーが企業に対して抱く同一性や帰属性の測定である。Bergami and Bagozzi（2010）は，回答者が自分と企業が重なり合うことを口頭と絵で示すことができる2つの尺度で実証検定を行い，いずれの方法も信頼性があり非常に使いやすく妥当なものであることを確認した。1つの尺度は，「この企業の価値観は私の価値観と一致する」という文章についてどの程度賛同するか，あるいはしないかを口頭で答えてもらうものである。もう1つは，ベン図を用いて2つの円の重なり具合を絵で示してもらうものであり，回答者は自分と企業の価値観の重なり具合をもっともよく表す絵を選択する。この2つのスコアを合算することにより，ステイクホルダーが認識する自分と企業との一体感を測定する。

　2つ目の方法は信頼の測定である。価値観が一致すると感じる企業に対して必ずしもすべてのステイクホルダーが親近感を抱くわけではないが，統合は企業の信頼性や誠実性に対するステイクホルダーの確信，すなわち信頼感が高まった結果であるとも言える（Morgan and Hunt, 1994）。信頼はステイクホル

ダーがサステナビリティに反応する上で欠かせないものであり，企業との一体感と比べるとサステナビリティ情報に左右されやすいが時間をかけて安定化していく傾向がある。したがって，統合を定量化するに当っては，企業との一体感の度合いと企業への信頼の両方を測定することが大切である。

4-1-2. ステイクホルダー・ルートの望ましい結果

　これまで議論してきたように，サステナビリティへの取り組みが生みだす価値は，多くがステイクホルダーの反応から生まれる。サステナビリティ活動を評価する際にはその価値を定量化すること，可能であればステイクホルダーの行動を金銭的に換算することが重要である。経営者はサステナビリティによって顧客のロイヤリティが高まることを期待し，業界におけるマーケットリーダーを目指している。ここで最も重要なのは，企業や経営スタイルによって異なる望ましい結果の選び方ではなく，サステナビリティ活動を通してステイクホルダーの行動に働きかける目標の設定の仕方である。

　サステナビリティ価値を測定する最も良い方法は，消費者，従業員，投資家などさまざまな形で企業にかかわる人々の行動を追跡することである。サステナビリティに関する意思決定において常に複雑な課題に直面している経営者にとって，単にサステナビリティ活動全体と業績変化の相関が分かるだけでは全く役に立たない。ステイクホルダーの行動のパターンを把握せず，サステナビリティに関する支出と業績の関係の詳細が明らかでない状態で意思決定を行うと，必ずしも望ましくない結果を，最悪の場合にはまずい結果を招くことになる。ステイクホルダー個々の行動は業績改善に向けてまとまっていくこともあるが，全体としてしか測定していないと，好意的に反応したステイクホルダーとそうでないステイクホルダーを把握することが難しい。したがって，ステイクホルダー・セグメントの中で観察しうる重要な行動をいくつか選び出し，測定することが望ましい。たとえば，消費者の購買行動は目に見える価値であり，客観的に観察することが可能である。行動を測定することが難しい場合には，購買意図を測定すればよい。入社を希望する人物の行動も同じく測定可能なことと似ている。

　この研究フレームワークにおける本質は，サステナビリティ価値には社会的

価値と財務的価値の両方が含まれるということである。ステイクホルダーの社会的な行動は，サステナビリティ活動を行う場合に重要な要素であるため，選定したコーズに対するステイクホルダーの行動を可能な限り常時追跡するべきである。ステイクホルダーの反応を測定する方法にはアンケートやインタビューがあり，これらにより社会的価値を測ることができる。例えば企業のサステナビリティ活動によって，そのサステナビリティ・プログラムと同じ社会的課題に対する寄付やボランティア，あるいはチャリティに対する積極的な口コミが増えたかどうか，またどのようにそれらを促進したか質問するのである。参考になる先行研究として，自分が購入している商品を販売する企業が支援するのと同じチャリティ活動に消費者が行う寄付について検証・測定したものがある。そこでは，消費者が企業に対して感じる統合が強くなるほど，対象となるチャリティへの寄付が増えることが報告されている（Lichtenstein et al., 2004）。

4-2. サステナビリティと企業価値の関係を発見する方法

　前節で示した測定方法は重要であるが，これだけでは経営者にとってサステナビリティが実際にどう機能しているのか分からない。ステイクホルダー中心主義であろうとすれば，ステイクホルダーの認識と行動の関係を深く理解することが必要である。

　サステナビリティと企業価値の関係を立証するには，研究フレームワークにおける1つの変数の変化が他の変数の変化と関係があるかどうか検証することになる。こうした関係が事例証拠や経営者の経験によって示されることがあるが，サステナビリティ活動に対するステイクホルダーの反応は複雑かつ心理的なものであるため，事例証拠だけでは効果的な意思決定の根拠にはならない。経営者は，研究フレームワークの変数間の関連やステイクホルダーの行動理由について注意深く分析を行い，意思決定を行う必要がある。

　研究フレームワークにおいて，サステナビリティと企業価値の関係を立証するのに欠かせない2つのステップがある。1つ目は，研究プロセスを確立させるようデータを収集する適切な研究デザインを行うことである。心理プロセス

を明らかにする上で，一般的にアンケート，室内実験，フィールド実験によって得られた1次データが不可欠であるが，店舗のレジなどで得られた2次データが基礎資料となることもある。研究デザインの一例として，われわれは企業から大学に対して行われた高額寄付のインパクトを評価する前後比較実験の方法を用いた。寄付前と寄付後に，ランダム化比較試験法によってステイクホルダーのデータを集めたところ，企業が寄付を行ったことを認識したステイクホルダーは認識していないステイクホルダーと比べてその企業に対する求職，商品購買，株式投資の意図が大きく高まることが明らかになった（認識不足は寄付に関する発表の前に調査を行ったことによるのか，単にキャンペーン情報を得ていなかったことによる）。この実験によって，寄付プログラムが企業に対するステイクホルダーの支持行動を活発化させたことが立証された。

　サステナビリティと企業価値の関係を立証するための2つ目のステップは，適切な統計技法を用いることである。こうした技法は高度に進化したものから相関係数の計算のように基本的なものまであるが，ほとんどの技法は単一／複数の変数の増加／減少が他の変数の規則的変化を生むのか，少なくとも関連があるのか検定を行うために設計されている。たとえば，われわれがShuili Duと行った共同研究では，ストーニーフィールドのヨーグルトの消費者の理解が深まることと統合が進むことには有意な関係が見られた（Du et al., 2007）。理解と統合の両方とも7段階尺度を用いて測定を行った（7は最も肯定的な理解と最高レベルの統合を意味する）。商品の品質に関する意見など，他の因子は全て同一とした場合，7段階尺度で理解が1段階進むごとに統合が0.3増加するという関連性がみられた。したがって，目標とする統合レベルに到達するためには企業はどれほど理解の改善を図る必要があるのかわかる。

　適切に上記2つの段階をふむことにより，経営者は望ましい行動を促進する要因を明らかにすること，更なる影響要因を特定することが可能になる。

4-2-1. 企業価値を高める要因を明らかにする方法

　企業価値を高めるのは，理解，有用性，統合の3つのUである。サステナビリティを企業価値に結びつけるには，3つのUと望ましい結果の関係性から考えていくことである。その最もわかりやすい方法は，個々のステイクホル

ダーのサンプルを対象に，3つのUと望ましい結果の両方を測定することである。例えば，あるステイクホルダーの企業に対する3つのUと，望ましい行動（購買意図や支払意欲など）を記録する。3つのUと結果の関係は，標準的な統計手法（例えば重回帰分析）によって分析される。

未だ実施されたことのないサステナビリティ概念を調べるには，どのように実験を行うかが重要であろう。例えば，サステナビリティの課題，プログラムの実施，資源のいずれかあるいは3つとも異なるプログラムの中から研究者が1つ選び，別々のステイクホルダーに説明を行う。ある説明を受けたステイクホルダーの回答と，他の説明を受けたステイクホルダーの回答を比較するという実験である。例えば，オーガニック原料を使用したコットンTシャツの価値を調べた実験がある。Trudel and Cotte（2009）は，100％オーガニックコットンの場合，50％の場合，20％の場合，その情報を示さない場合，サステナブルな生産ではない場合，それぞれに潜在的に消費者が存在することを示した。サステナブルな生産によるTシャツと比べて，そうでないTシャツに対しては消費者の購買意図が明確に減るという結果が表れた。興味深いことに，オーガニック原料の使用度合いによって消費者の購買意図が変わるわけではなかった。100％オーガニックコットンTシャツに対する購買意図は25％オーガニックコットンのそれよりも高かったものの，その差はわずかであった。消費者が企業のサステナビリティ活動を知れば企業価値を生むことは明らかになったが，購買意図を測定するこの実験方法だけでは，100％オーガニックであることが企業価値を生み出すとは言い切れない。

4-2-2. 他の影響要因を特定する方法

ステイクホルダー中心主義の企業は，状況に応じてサステナビリティへの取り組みをうまく行っている。サステナビリティ活動の効果を調整する際には（その調整に関して後述）2つのレベルの要因—ステイクホルダーに関する要因と企業に関する要因—を測定し，追跡し，サステナビリティに関する意思決定に組み込む必要がある。

・**ステイクホルダーに関する要因**

ステイクホルダーに関する要因の測定には，しばしばアンケートが用いられ

る。われわれは質的データと量的データを用い，統計と調査を組み合せて行った。例えば，企業のダイバーシティプログラムに関する研究においてサステナビリティ課題のリストを作成し，各課題をどの程度支持するか回答者に尋ねることによって（7段階尺度で1＝まったく支持しない，7＝積極的に支持する），ステイクホルダーの関心のレベルを測定した。課題リストには，平等な雇用機会，同性愛者のための特別な雇用支援，障害者のための特別な雇用支援，少数民族のための特別な教育機会など10項目がある。回答結果の平均値を算出し，ダイバーシティというサステナビリティ課題に対する関心レベルという1つの指標とした。

質的ケーススタディでは，世界各地に勤務するFortune500の企業の従業員を対象に，サステナビリティ・プログラムの有用性について質問した。そしてグループごとに聞き取り調査から各地の従業員の回答内容を比較した結果，有用性は勤務地によって大きく異なることが明らかになった。本社勤務の従業員は，新しいスキルを身につける機会となるなど自らの成長のために役立ったときにサステナビリティを有用なものと評価していたが，自社が現地住民から批判を受ける状況にある海外支社勤務の従業員は，批判をかわすのに役立ったときに最もサステナビリティ・プログラムを有用であると評価していた。

量的ケーススタディでは，前述した口腔ケアプログラムがもたらした便益について検討を行った。このプログラムに参加した人々は参加していない人々と比較して理解度が深まり，プログラム実施企業の歯磨き粉ブランドを購入する傾向が強まった。しかしアメリカでの生活が短い参加者（文化順応度は低い）と，長い参加者（文化順応度は高い）の回答を比較したところ，その差異はさらに明確になった。アメリカでの生活期間が短いグループはアメリカ文化に順応しようとする意志が強いため，期間の長いグループよりも当該企業の商品に対する強い理解と購買意図を示している。サステナビリティ・プログラムは，すべての参加者を対象として企業に対する関心を高めようとしているのであるが，とくにアメリカでの生活期間の短い人々に対して有効であることが，この研究の結果明らかになった。

・企業に関する要因

　われわれの研究では，企業価値は企業によって異なる形でつくられ，同じサステナビリティ活動でも企業によって非常に異なる結果を生む可能性があることが明らかになった。こうしたことは産業，企業規模，あるいは他の客観的相違によって起こることもあるが，ステイクホルダーがどのくらい企業のことを知っているかによって起こることが多い。例えば，サステナビリティに対するステイクホルダーの反応は，彼らがもつ企業に対する知識と能力によって異なる。Becker-Olsen et al.（2006）は，企業の本業とサステナビリティ活動の一致度がもたらす影響を検証している。この研究では回答者をランダムに指定し，あるサステナビリティ活動について，その活動と本業が近い企業が実施しているのか，あるいは本業とは全く異なる企業が実施しているのかという観点から評価してもらった。その結果，サステナビリティ活動と本業とが遠い企業に対しては，近い企業と比べて回答者が懐疑心をもつことが明らかになった。

4-3. CSRプログラムを調整する方法とステイクホルダーとの関係

　サステナビリティを戦略的にマネジメントするに当っては，絶え間なく改善していくよう定期的にサステナビリティ・プログラムを調整するプロセスが必要である。サステナビリティと企業価値の関係を定量化し明らかにすることが重要であり，そのことでステイクホルダーとの関係を最大限強固にする知見が得られる。サステナビリティ活動を調整するためこの知見を適切に活用することでサステナビリティ活動に対する理解を深め，ステイクホルダーのニーズを満たし，企業価値を最大化することができる。

　具体的には，この知見を既存のCSRプログラム，これから展開するCSRプログラム，社内他部署の3つに活用し，ステイクホルダーとの関係を向上させていくことによって便益を得ることができる。後者の2つは，現在行っているCSRプログラムを評価し，その情報を関係者と共有していくことから生まれてこよう。

4-3-1. ダッシュボード：知見をサステナビリティ活動に活かす方法

　サステナビリティへの取り組みとステイクホルダーとの関係を調整していく

にあたり重要な第一歩は，各プログラムに関する全ての知識を継続的に集約していくことである。先進的企業は，経営者が一度にすべての関連情報をみることができるダッシュボードのようなシステムをつくっている。財務，顧客，業務プロセス，学習と成長によって業績評価を行う有名な「バランス・スコアカード」(Kaplan and Norton, 1993) のように，本研究フレームワークにおける各段階の主要な指標に基づき CSR ダッシュボードをつくることができる。ここには，理解，有用性，統合，および2種類の企業価値―財務的価値と社会的価値―が含まれる。ダッシュボードは，これらを評価することによって現在の業績をみるだけではなく，次期の目標を示すこともできる。

　ダッシュボードは，サステナビリティ・プログラムを体系的かつ継続的に改善していくために活用することができる。機能しているものとしていないものを把握するには時間がかかるが，CSR に対するステイクホルダーの反応もまた徐々に発展していくものである。サステナビリティへの取り組みを導入してから数週間あるいは数カ月もの間その成果が明らかにならないことはよくある。そこですべてのサステナビリティ活動が最大の成果をあげられるよう，ダッシュボードを使って目に見える形で共有された目標に向かって進捗を追跡していくとよい。

　サステナビリティ戦略を評価するにはギャップ分析が極めて有効である。目標を達成できない時，目標と実際の結果の間にギャップがあることは明らかであり，なぜそのプログラムが成果を出せないのか理解，有用性，統合それぞれに内在する要因まで掘り下げて検討することができる。多くの場合，その情報に基づき理解を促進するようコミュニケーションを改善したり，有用性を高める機能を付加したりすることによって，サステナビリティ・プログラムを再調整することができる。もし再調整されたプログラムが実現可能でない場合やコストがかかりすぎる場合は，断念することが最良の決断になることもある。

4-3-2. 知見の活用によるサステナビリティ活動ポートフォリオの最適化

　サステナビリティ戦略の策定にあたり，ステイクホルダーとの関係を可能な限り強固にしていく知識を得るには，定量的分析とインタビュー調査を組み合わせていくことが重要である。先進的な経営者は，ステイクホルダーとの関係

を構築していくことを通して，財務的目標と社会的目標を達成するようサステナビリティを捉えている。したがって CSR プログラムを単独の取り組みではなく，様々な取り組みの組み合わせとして捉えるべきであり，他の CSR に関連する部署もそれらの知識を共有することで，より活用することができる。

　このような組織内での共有化の必要性が明らかであるにもかかわらず，われわれの経験上，マーケティング，営業，研究開発，広報の各部署が連携せずサステナビリティ活動を実施することが一般的である。したがって，意思決定を行う担当者全員が，サステナビリティ評価ダッシュボードのようなツールを使えるようにしておく必要がある。イギリスのマークス＆スペンサーは，全ての部署と従業員がサステナビリティ戦略に関わっている企業の一例である。同社は，バリューチェーン全体において 180 にのぼる社会的・環境的目標について成果測定を行い，その達成状況をロンドン本社からイントラネットで全従業員に向けてリアルタイム発信している（Hollender and Breen, 2010）。

　サステナビリティ担当者間で知識を共有することが求められるのには 2 つの理由がある。1 つは，情報発信から反応まで多くの段階をふむ本研究のフレームワークは複雑であり，サステナビリティ活動担当者が，企業のサステナビリティ戦略全体に対するステイクホルダーの反応を正確につかむことは容易ではない。サステナビリティ担当部署が他の取り組みによって得られた最新の知識も入手できれば，より効果的なサステナビリティ戦略を策定することができるであろう。そこでより上位の管理者層は，できるだけ多くのサステナビリティ活動関連担当者が，あらゆる関連情報を考慮に入れて調整を行っていけるよう，ステイクホルダーの反応に関する最新の知識を入手できる措置を講じる必要がある。

　もう 1 つの理由は，多くのステイクホルダーは同時に数えきれないほどのサステナビリティ・プログラムに接しているということである。彼らは企業の活動に一貫性を求めており，サステナビリティ・プログラム間の調整不足は混乱を招き，ステイクホルダーに懐疑心をもたせてしまいかねない。

　このように，サステナビリティ戦略はステイクホルダーに価値を提供する一つ一つの取り組みを組み合わせたものとして捉えることができる。先進的企業

は，知識の共有を進めることによってこれらの役割をマネジメントし，サステナビリティ・プログラム間の相乗効果を生み出している。

5. 結論

　本稿では，サステナビリティ経営が非常に複雑でさまざまな要因に左右される取り組みであることを明らかにした。このような経営を行っていくにはアートと科学の両方が必要である。ステイクホルダー中心主義の企業になるためには，サステナビリティ戦略においてステイクホルダーからのインプットと参画（共創）を求め，サステナビリティ活動に関する正しい情報を提供し（コミュニケーション），企業価値を最大化するために継続的に活動を追跡および調節（調整）することが要求される。したがって，サステナビリティ戦略は目的というよりプロセスとして理解することが望ましい。

　顧客中心主義の研究者は，顧客中心主義の企業は顧客が何に価値を置いているかということのみならず，自社にとって顧客がどれほど価値があるかということについて明らかにしている。こうした企業は，単に顧客に対して友好的であるだけではなく，長期的に多くの利益をもたらす顧客がもつ基本的ニーズに応えるよう努めている。ステイクホルダー中心主義の企業となるためにサステナビリティに取り組むことは，これとまったく同じことである。経営者は，自社にとって長期的に最も重要なステイクホルダーを特定し，彼らとの長期的関係を構築していく方法をみつけなければならない。3つのU（理解，有用性，統合）のアプローチは，そのために設計されたものであり，サステナビリティ活動に対するステイクホルダーの解釈，評価，行動の心理的プロセスについて1つの視座を提供するものである。

本稿は下記書籍に基づいている。詳しくは本書を参照されたい。
Bhattacharya, CB, Sen, S. & Korschun, D.（2011）*Leveraging Corporate Responsibility: The Stakeholder Route to Maximizing Business and Social Value*, Cambridge University Press.

< 参考文献 >

Becker-Olsen, K. L., Cudmore, B. A. and Hill, R. P. (2006) 'The impact of perceived corporate social responsibility on consumer behavior', *Journal of Business Research*, Vol. 59, No. 1, pp. 46-53.

Bergami, M. and Bagozzi, R. P. (2010) 'Antecedents and Consequences of Organizational Identification and the Nomological Validity of the Bergami-Bagozzi-Scale', Working paper, Marketing Department, Jones Graduate School of Management, Rice University.

Bhattacharya, CB, Sen, S. and Korschun, D. (2008) 'Using Corporate Social Responsibility to Win the War for Talent', *Sloan Management Review*, Vol. 49, No. 2, pp. 37-44.

Bonini, S., Koller. T. M. and Mirvis Ph. (2009) 'Valuing Social Responsibility Programs', *McKinsey on Finance*, Vol. 32, pp. 11-18.

Cohen, C. and Prusak, L. (2001) *In Good Company: How Social Capital Makes Organizations Work*, Boston: Harvard Business School Press.

Cone (2007) 'Cause evolution survey', Available at: http://www.conecomm.com/2007-cause-evolution-and-environmental-survey, Accessed January 12, 2015.

Dawkins, J. (2004) 'Corporate Responsibility: The Communication Challenge,' *Journal of Communication Challenge*, Vol. 9, No. 2, pp. 108-119.

Dwyer, F. R., Schurr, P. H. and Oh, S. (1987) 'Developing Buyer-Seller Relationships', *Journal of Marketing*, Vol. 51, No. 2, pp. 11-27.

Du, S., Bhattacharya, CB and Sen, S. (2007) 'Reaping Relationship Rewards from Corporate Social Responsibility: The Role of Competitive Positioning', *International Journal of Research in Marketing*, Vol. 24, No. 3, pp. 224-241.

Fombrun, C. and Shanley, M. (1990) 'What's in a Name? Reputation Building and Corporate Strategy,' *Academy of Management Journal*, Vol. 33, No. 2, pp. 233-258.

Greening, D. W. and Turban, D. B. (2000) 'Corporate Social Performance as a Competitive Advantage in Attracting a Quality Workforce,' *Business & Society*, Vol. 39, No. 3, pp. 254-280.

Haley, E. (1996) 'Exploring the Construct of Organization as Source: Consumers' Understanding of Organizational Sponsorship of Advocacy Advertising,' *Journal of Advertising*, Vol. 25, No. 2, pp. 19-36.

Hill, R. P., Ainscough, T., Shank, T. and Manullang, D. (2007) 'Corporate Social Responsibility and Socially Responsible Investing: A Global Perspective', *Journal of Business Ethics*, Vol. 70, No. 2, pp. 165-174.

Hollender J. and Breen, B. (2010) *The Responsibility Revolution*, San Francisco: Wiley and see also, http://plana.marksandspencer.com/about, accessed January 12, 2015.

〈翻訳〉1 サステナビリティ経営を成功させるステイクホルダー・ルート 71

Kaplan, R. S. and Norton, D. P. (1993) 'Putting the Balanced Scorecard to Work', *Harvard Business Review*, Vol. 71, No. 5, pp. 134-147.

Korschun, D. and Du, S. (2013) 'How Virtual Corporate Social Responsibility Dialogs Generate Value: A Framework and Propositions', *Journal of Business Research*, Vol. 66, No. 9, pp. 1494-1504.

KPMG (2008) 'KPMG International Survey of Corporate Responsibility Reporting', Available at: http://www.kpmg.com/EU/en/Documents/KPMG_International_survey_Corporate_responsibility_Survey_Reporting_2008.pdf, Accessed January 12, 2015.

Lawrence, A. T. (2010) 'Managing Disputes with Nonmarket Stakeholders: Wage a Fight, Wait, Withdraw, or Work It Out?', *California Management Review*, Vol. 53, No. 1, pp. 90-113.

Lichtenstein, D. R., Drumwright, M. E. and Braig, B. M. (2004) 'The Effect of Corporate Social Responsibility on Customer Donations to Corporate-Supported Nonprofits', *Journal of Marketing*, Vol. 68, No. 4, pp. 16-32.

Menon, S. and Kahn, B. E. (2003) 'Corporate Sponsorships of Philanthropic Activities: When Do They Impact Perception of Sponsor Brand?', *Journal of Consumer Psychology*, Vol. 13, No. 3, pp. 316-327.

Morgan, R. M. and Hunt, S. D. (1994) 'The Commitment-Trust Theory of Relationship Marketing', *Journal of Marketing*, Vol. 58, No. 3, pp. 20-38.

Petty, R. E., Cacioppo, J. T. and Goldman, R. (1981) 'Personal Involvement as a Determinant of Argument-Based Persuasion', *Journal of Personality and Social Psychology*, Vol. 41, No.5, pp. 847-855.

Rosen, S., Irmak, C. and Jayachandran, S. (2009) 'Value Co-Creation in Cause-Related Marketing: How Letting Consumers Choose the Cause Enhances Consumer Support', Working Paper, Moore School of Business.

Sen, S. and Bhattacharya, CB. (2001) 'Does Doing Good Always Lead to Doing Better? Consumer Reactions to Corporate Social Responsibility', *Journal of Marketing Research*, Vol. 38, No. 2, pp. 43-62.

——, —— and Korschun, D. (2006) 'The Role of Corporate Social Responsibility in Strengthening Multiple Stakeholder Relationships: A Field Experiment', *Journal of the Academy of Marketing Science*, Vol. 34, No. 2, pp. 158-166.

——, Du, S. and Bhattacharya, CB. (2009) 'Building Relationships through Corporate Social Responsibility', in MacInnis, D. J., Park C. W. and Priester, J. R. (Eds.) *Handbook of Brand Relationships*, M. E. Sharpe.

Simmons C. J. and Becker-Olsen, K. L. (2006) 'Achieving Marketing Objectives through

Social Sponsorships', *Journal of Marketing*, Vol. 70, No. 4, pp. 154-169.
Target (2008) 2007 Corporate Responsibility Report, Available at: https://corporate.target.com/_ui/pdf/2007_full_report.pdf, Accessed January 12, 2015.
The Harris Poll Reputation Quotient (2012) Available at: http://www.harrisinteractive.com/Products/ReputationQuotient.aspx, Accessed January 12, 2015.
Trudel, R. and Cotte, J. (2009) 'Does It Pay To Be Good?', *MIT Sloan Management Review*, Vol. 50, No. 2, pp. 61-68.
Yoon, Y., Gürhan-Canli, Z. and Schwarz, N. (2006) 'The effect of corporate social responsibility (CSR) activities on companies with bad reputations', *Journal of Consumer Psychology*, Vol. 16, Issue 4, pp. 377-390
Whole Foods Market. (2012) Available at: http://www.wholefoodsmarket.com/value, Accessed January 12, 2015.

2 Strategy and Sustainability
——Changing Conceptions to Enable a Sustainable Direction

Nick Barter
Associate Professor, Griffith University, Australia

[Abstract]
This paper explores two key areas before offering a framework to enable corporate strategists to realize sustainable outcomes. To do this the paper first outlines how sustainable development is actually a call for a change in our conceptions away from dualistic to monistic understandings. Second it discusses the basic components of strategy using bestselling strategy textbooks as reference sources. The paper then brings these two discussion areas together to realize a conceptual framework for the strategist that does not enable dualistic conceptions, but rather monistic conceptions that will in turn enable moves to more sustainable outcomes. As this paper is conceptual and challenging, its power lies in what it suggests rather than what it proves (Fiol, 1989).

Key Words : Strategy, sustainability, new conceptual framework, monistic, dualistic

"It is quite simply wrong to regard action on the psyche, the socius, and the environment as separate...[to present these areas as if they are separate is]...acquiescing to a general infantilisation of opinion" (Guattari, 1989, pp. 134)

1. Introduction

This paper builds upon a presentation made at the Japan Forum of Business and Society's fourth annual conference in September, 2014. That presentation explored understandings of strategy and sustainability as separate concepts and then brought them together to offer a challenging, yet simple, conceptual framework for realising sustainable strategies. That simple framework rests upon the notion that to enable organisations to realise sustainable outcomes, separatist conceptions need to be removed and simple questions asked to test the organisation's strategy. One part of the framework's test questions puts the individual—the 'I' of the decision maker—as the locus of concern. This avoids separatist conceptions that might enable the decision maker to make concerns about the organisation's strategies be about impact on an abstract other, but rather shifts the focus to the questioner. When the concern is about the abstract other, the burden of responsibility can easily be shifted; however, the use of a personal question ensures the locus of responsibility and impact is not ameliorated. The 'I' personalises, and in so doing puts the decision maker into a more difficult bind regarding justification of actions. In this regard, one of the central arguments in this paper is that in order to enable sustainable outcomes, a strategy should have asked of it, by the strategy decision makers and stakeholders, simple personal questions such as: as a result of this strategy, will the air I breathe be cleaner? Or alternatively, would I want to breathe the gaseous outputs of this process? Such simple questions are easily conceptualised, understood and answered by any individual. Furthermore, if the answer is 'no' then the decision maker has to clarify how they justify such a strategy, especially within our current epoch, the Anthropocene (Crutzen, 2006), where humans are recognised as the primary shapers of the world around us. Thus, a negative answer challenges an individual decision maker to justify or develop new strategies in order to realise a positive response. Asking about the quality of air is only one aspect of the framework, and the paper develops a suite of these and other simple questions across the classic domains of environment, society and economy to develop a more complete framework. It should be noted that the domains of environment, society and economy are challenged by the arguments in this paper. However, such domains enable analysis and are thus useful facilitators of action; in short, they enable a checklist and comprehensive approach to be taken—a necessary function of any framework for the organisational decision maker.

To develop the framework and the arguments supporting it, the paper begins by first outlining an understanding of sustainable development through its history and key implications. Second, it outlines the basic understandings of strategy and what it is at a fundamental level, building this understanding via a simple content analysis of definitions of strategy in popular textbooks. These two discussions are foundational and synthesised in the third part of the paper, where the understandings of strategy and sustainability are woven to realise the conceptual framework for developing organisational strategies that can have sustainable outcomes. This third part of the paper also develops a framework that is comprehensive and applicable across multiple organisations and their products and services. The fourth and final part of the paper ends with some reflections and considerations.

To close this introduction, it should be noted by the reader that this paper is offering a conceptual framework that is built from deducing the logical implications of key arguments; however, that framework has not been tested. As such, the value of this paper is not what it 'proves', but rather what it suggests (Fiol, 1989). Finally, the opening quote of this paper, from Guattari (1989), outlines a challenging but readily known understanding that we live in a borderless world and thus our "awareness of the mutual embedding of humans and the rest of nature…[is]…so obvious that many of us take this special connection for granted" (Starik & Kanashiro, 2013, p. 8). In this regard, the framework offered by this paper is arguably simple and logical, yet our realisation of this obvious simplicity is clouded because of our ingrained perspectives.

2. Understanding Sustainable Development

The roots of sustainable development can be found within the environmental movement of the 1960s and the publication of key texts (Carruthers, 2001; Tulloch, 2013; Tulloch & Neilson, 2014), such as Silent Spring (Carson, 1962), The Population Bomb (Ehrlich, 1971) and The Limits to Growth (Meadows, et al; 1972). It is argued that these texts, when coupled to Hardin's (1974) lifeboat ethic and the concept of carrying capacity, helped inform the discourse of the time, with that discourse entering the mainstream via *Our Common Future* (WCED), the 1987 United Nations publication (Carruthers, 2001; Shrivastava & Hart, 1994; Steer & Wade-Gery, 1993; Tulloch, 2013; Tulloch & Neilson, 2014; Yates, 2012).

Our Common Future (1987) defined sustainable development as "development which meets the needs of the present without compromising the ability of future generations to meet their own needs" (WCED, 1987, p. 8). Although this definition suffers critiques, not least that it is a compromise aimed at dissolving historical conflicts between ecological and economic concerns (for example see Banerjee, 2003; Carruthers, 2001; Tulloch, 2013; Tulloch & Neilson, 2014), it is generally accepted that sustainable development is now the dominant global discourse regarding ecological concerns (Carruthers, 2001; Tulloch, 2013). It is also an idea that is increasingly important for organisations and business leaders to embrace (for example see; Hopkins, 2009; Kiron et al., 2012; Kiron et al., 2013), given it is argued that we "live in an organizational world in which organizations are the means through which interests are realized" (Egri & Pinfield, 1999, pp. 225). The embrace of sustainability by organisations implies that organisational leaders, and in turn corporate strategists, are key, especially as these individuals are critical in marshalling the resources and setting the direction of organisations. The importance of corporate strategists in realising sustainable outcomes is recognised in the United Nations 2012 update to *Our Common Future* (1987), *Resilient People, Resilient Planet* (2012). The report identifies "corporate strategists" (UNSGHLPGS, 2012, p. 22) as key actors who "have more opportunity than ever to pick and choose from the best practices and resources ... combine them in new and previously unforeseen ways...[and thus help]... to drive sustainable development" (UNSGHLPGS, 2012, p. 22). While such recognition potentially highlights how the concept of sustainable development has been captured by commerce and "morphed to accommodate neoliberal assumptions" (Tulloch, 2013, p. 109) away from its radical roots, the challenge remains, to develop a framework to enable sustainable development that corporate strategists can utilise and that reflects the central thesis of the concept (for example see; March & Simon, 1958; Simon, 1947; Lakoff, 2010).

The central thesis of sustainable development is present and future generations and their development-in short human progress and survival. In this regard, it should not be lost that sustainable development is a concept that is written by humans and for humans, and thus it is a human-centric concept.

Moving beyond this base understanding, a key challenge regarding the realisation of sustainable development is our understanding of the environment and whether we consider it as something that is separate and out there, something that has equivalence with the economy or not, and/or something that is inseparable from ourselves. Underlying this is the difference between dualistic and monistic understandings and, with regard to how the environment is conceptualized. It is argued that the realisation of sustainable outcomes requires a move away from the concept of dualism. *Our Common Future* (1987) argues that "the environment does not exist as a sphere separate from human actions, ambitions, and needs, and attempts to defend it in isolation from human concerns have given the very word environment a connotation of naivety" (WCED, 1987, p. xi). This is a direct statement on how the dualistic consideration of the environment as a separate sphere of concern to ourselves and our actions is not appropriate for the realisation of sustainable outcomes. It implies a move away from considering the environment as a thing that is bounded, separate and external to humanity, to a more monistic understanding in which individuals recognise that they are their surroundings (Tulloch, 2013; citing Suzuki, 2008).

To build the case, *Our Common Future* (1987) argues that humans have historically understood the planet as "a large world in which human activities and their effects were neatly compartmentalised within nations, within sectors (energy, agriculture, trade) and within broad areas of concern (environmental, economic, social)" (WCED, 1987, p. 4). Such an understanding conceptualises the world as consisting of separate domains of, for example, environment, society and economy. However, this ultimately reinforces a false dichotomy that there are two separate categories in the world; humans and everything else (nature) (for further discussion see; Castree, 2002; Latour, 1999a, 1999b; Newton, 2002). This is counter to the "real world of interlocked economic and ecological systems...[that]... will not change" (WCED, 1987, p. 9). Thus if the 'real' world will not change, the challenge is whether humanity's understandings and our "policies and institutions" (WCED, 1987, p. 9) can, lest sustainable outcomes be forfeited.

Taken together, the call for sustainable development is a call for a move away from a fractured epistemology that separates humans from everything else. Thus, the call for sustainable development is also a call for new knowledge structures and systems that do not separate, are not fractured. Humans need to embrace an understanding that the environment is not separate and out there; rather, it surrounds and is entwined with us (Ingold, 2011). In the terms of Starik and Kanashiro (2013), we need to recognise the obvious, the "mutual embedding of humans and the rest of nature" (p. 8). And even though this mutual embedding may be "so obvious that many of us take this special connection for granted" (Starik & Kanashiro, 2013, p. 8), the challenge is that "all human organisations are embedded within the natural environment, and … all of those which have human managers and other employees, also contain the natural environment inside of their respective biophysical bodies" (Starik & Kanashiro, 2013, p. 9). As such there is no separation, and to reinforce an earlier point, sustainable development is not so much about saving the environment as it is about humans and how we wish to live.

Sustainable development has been defined as "development which meets the needs of the present without compromising the ability of future generations to meet their own needs" (WCED, 1987, p. 8). Without wishing to get lost in critical debates regarding such a definition (for example see Banerjee, 2003; Carruthers, 2001; Tulloch, 2013; Tulloch & Neilson, 2014), what is evident from it is that it concerns humans, our development and a time dimension relating the present to the future. It has also been defined as a "relationship between dynamic human economic systems and larger dynamic, but normally slower-changing ecological systems, in which (a) human life can continue indefinitely, (b) human individuals can flourish, and (c) human cultures can develop; but in which effects of human activities remain within bounds, so as not to destroy the diversity, complexity, and function of the ecological life support system" (Costanza, Daly, & Bartholomew, 1991, p.8). Further and building on prior studies and a content analysis of different definitions of sustainable development, Gladwin, Kennelly, and Krause (1995) defined it as "a process of achieving human development in an inclusive, connected, equitable, prudent, and secure manner" (p. 878).

Pulling the components of these definitions together, sustainable development concerns humans, our survival, continuation, development and wellbeing; it is a purpose or path for humans to follow—it is not about saving the environment. Furthermore, that path— as outlined earlier—is enabled by organisations and thus corporate strategists operating with an epistemology that is not fractured and separatist.

3. Understanding Strategy

Understanding strategy can be a difficult process, if the provocation in the title of Whittington's (1993) text 'What is Strategy—and does it matter?' is accepted. However, there are also a plethora of strategy textbooks on the market indicating the opposite, particularly as these texts have been likened to recipe books (for example see; Spender, 1989; Whittington, 1993) offering the reader the delightful result of a successful strategy (Spender, 1989; Whittington, 1993). The definitions of strategy found in these books are a useful field from which to distil the key elements of the concept; however, given the vast number of them on the market, there is a challenge present in which to focus on. Identifying bestsellers from a more manageable list of textbooks is important; in November 2011, publishing houses (for example; Wiley, Pearson, McGraw Hill and Cengage) were contacted for advice and bestseller rankings for sales of strategy/strategic management textbooks in the territories of the UK, Australia and the United States of America. In response to those requests, some of the publishers provided a list of best-selling titles, or in one case a list of titles with university course adoption figures alongside each title. In addition to the publishing houses, and as a way of triangulating the data obtained, international book sales data monitoring company Nielsen BookScan was contacted for sales data on strategy textbooks within the aforementioned territories. Also, as an additional way of triangulating the data, the University Co-operative Bookshop chain of Australia was contacted for strategic management textbook sales in the year to October 2011 (the most recent 12-month period they could provide at the time of the request). Within Australia, the University Co-Operative bookshop chain is the largest seller of textbooks and has a presence on virtually every Australian campus, thus its sales data is indicative of the Australian market. In sum, lists and rankings were obtained from a variety of sources, albeit no source provided data on the United States of America. Those lists resulted in a long list of 46 strategic management textbooks.

Reviewing the Australian sales volume data from the University Co-Operative Bookshop revealed that approximately 95% of strategy textbook sales were accounted for by just 10 titles. Given this, and if Australian data is considered an exemplar, the long list of 46 titles was reduced to a more manageable list of 23 titles for review. To create that shorter list of the bestselling textbooks in Australia and the United Kingdom, the following decision rules were applied. A textbook would be included in the final shortlist if: (1) it was ranked in the top ten of either an Australian or UK ranking according to a sales data monitoring organisation; (2) it was in the top 20 of an Australian and UK ranking according to a sales data monitoring organisation; (3) it was in the top ten of sales according to the University Co-Operative Bookshop of Australia; (4) it was included in the top ten of other lists that were provided on bestselling or course adopted textbooks by publishers. The final list of textbooks reviewed is listed below, along with their definitions of strategy.

Example definition from the textbooks define strategy as "the determination of the long-run goals and objectives of an enterprise, and the adoption of courses of action and the allocation of resources necessary for carrying out these goals" (Angwin, Cummings & Smith, 2008, p. 119); or how "organisations exist to fulfil a purpose and strategies are employed to ensure that the organisational purpose is realised" (De Wit & Meyer, 2010a, p. 6). Alternatively, how "strategic management allows an organisation to be more proactive than reactive in shaping its own future" (David, 2011, p. 48) and how "a firm's strategy is defined as its theory about how to gain competitive advantages. A good strategy is a strategy that actually generates such advantages" (Barney & Hesterly, 2010, p. 4).

Table 1 Strategy Textbook Titles

Strategy Textbook	Definition of Strategy
1. Angwin, D. Cummings, S. & Smith, C. (2008) *The strategy pathfinder: Core concepts and micro-cases,* Oxford: Blackwell Publishing.	strategy is "the determination of the long-run goals and objectives of an enterprise, and the adoption of courses of action and the allocation of resources necessary for carrying out these goals" (p. 119)
2. Barney, J. B. & Hesterly, W. S. (2010) *Strategic Management and competitive advantage: Concepts and Cases* (3rd ed.), New Jersey, USA: Prentice Hall.	"a firm's strategy is defined as its theory about how to gain competitive advantages. A good strategy is a strategy that actually generates such advantages" (p. 4) "The strategic management process is a sequential set of analyses and choices that can increase the likelihood that a firm will choose a good strategy; that is a strategy that generates competitive advantages" (p. 4) "In general, a firm has a competitive advantage when it is able to create more economic value than rival firms" (p. 10) "Economic value is simply the difference between the perceived benefits gained by a customer that purchases a firm's products or services and the full economic cost of these products or services" (p. 10)
3. Besanko, D., Dranove, D., Shanley, M and Schaefer, S. (2010) *Economics of strategy* (5th ed.), New York: John Wiley & Sons.	Conducts some simple content analysis of three definitions of strategy. The text outlines how "strategy has to do with the big decisions a business organization faces, the decisions that ultimately determine its success or failure" (p. 1). Further strategy "defines what kind of company it is or should be" (p. 1) such that strategy concerns the "firm's competitive persona, its collective understanding of how it is going to succeed within its competitive environment" (p. 1)
4. Capon, C. (2008) *Understanding Strategic Management,* Harlow: Pearson Education.	"the reader will be able to make up his or her own mind on what constitutes strategy" (p. 5) via reading the book. Outlines the advantage of strategy (p. 5)-involves

Strategy Textbook	Definition of Strategy
	the whole organisation, focuses on the relationship between the organisation and its environment, includes the managers and leaders, covers all activities, is central to creating competitive advantage, allow organisations to maximise competitive advantage and or survive
5. Carpenter, M. A. and Sanders, W. G. (2009) *Strategic management: A dynamic perspective concepts and cases* (2nd ed.), New Jersey, USA: Pearson, Prentice Hall.	"Strategy is the coordinated means by which an organisation pursues its goals and objectives" (p. 10)
6. David, F. R. (2011) *Strategic management: Concepts* (13th ed.), New Jersey, USA: Pearson Education.	"Strategies are the means by which long-term objectives will be achieved" (p. 45) "Strategic management allows an organisation to be more proactive than reactive in shaping its own future" (p. 48)
7. De Wit, B. & Meyer, R. (2010a) *Strategy process, content, context: An international perspective* (4th ed.), Mason USA: South Western Cengage Learning.	"in this book we will proceed with a very broad conception of strategy as 'a course of action for achieving an organisation's purpose" (p. 52) "Organisations exist to fulfil a purpose and strategies are employed to ensure that the organisational purpose is realised" (p. 6) "Strategic management is concerned with relating a firm to its environment in order to successfully meet long-term objectives" (p. 236)
8. De Wit, B. & Meyer, R. (2010b) *Strategy synthesis: Resolving strategy paradoxes to create competitive advantage* (3rd ed.), Andover, UK: Cengage Learning.	a broad definition is offered whereby "in this book we will proceed with a very broad conception of strategy as 'a course of action for achieving an organisation's purpose" (p. 257) "The broader set of fundamental principles giving direction to strategic decision making, of which organisational purpose is the central element, is referred to as the corporate mission" (p. 257)
9. Dess, G. G., Lumpkin, G. T., Eisner, A. B. & McNamara, G. (2010) *Strategic management texts and cases* (5th ed.), New York, USA: Mc-Graw-Hill Irwin.	"Strategy the ideas, decisions, and actions that enable a firm to succeed" (p. 9)

2 Strategy and Sustainability

	Strategy Textbook	Definition of Strategy
10.	Grant, R. M. (2010) *Contemporary strategy analysis* (7th ed.), West Sussex: Wiley & Sons.	"In its broadest sense, strategy is the means by which individuals or organisations achieve their objectives" (p. 16)
11.	Hanson, D., Hitt, M. A., Ireland, R. D. & Hoskisson, R. E. (2011) *Strategic management: Competitiveness and globalisation* (Asia–Pacific 4th ed.), Melbourne: Cengage Learning Australia.	"A strategy is an integrated and coordinated set of commitments and actions designed to exploit core competencies and gain competitive advantage" (p. 4)
		"A firm has a competitive advantage when it implements a strategy competitors are unable to or find too costly to try to imitate" (p. 5)
12.	Harrison, J. S. & St. John, C. H. (2010) *Foundations in strategic management* (5th ed.), Mason USA: South Western Cengage Learning.	"Strategic management is the process through which organisations analyse and learn from their internal and external environments, establish strategic direction, create strategies that are intended to help achieve established goals, and execute those strategies, all in an effort to satisfy key organisational stakeholders." (p. 4)
13.	Hill, C. W. L. & Jones, G. R. (2010) *Strategic management: An integrated approach* (9th ed.), Mason, USA: South Western Cengage Learning.	"a strategy is a set of related actions that managers take to increase their company's performance" (p. 3)
14.	Hill, C. W. L., & Jones G. R. (2012) *Essentials of strategic management* (3rd ed.), Mason USA: Southwestern Cengage Learning.	"A strategy is a set of actions that managers take to increase their company's performance relative to rivals" (p. 2)
15.	Hitt, M. A., Ireland, D. R. & Hoskisson, R. E. (2011) *Strategic management: Competitiveness and globalization: Concepts and cases* (9th ed.), Mason USA: South Western Cengage Learning.	"A strategy is an integrated and coordinated set of commitments and actions designed to exploit ore competencies and gain a competitive advantage" (p. 4)
		"A firm has a competitive advantage when it implements a strategy competitors are unable to duplicate of find too costly to try to imitate" (p. 4)
16.	Hubbard, G. & Beamish, P. (2011) *Strategic management: Thinking, analysis, action* (4th ed.), Frenchs Forest NSW: Pearson Australia.	Strategy is defined as "Those decisions that have high medium term to long term impact on the activities of the organisation, including the analysis leading to the resourcing and implementation of those decisions, to create value for key stakeholders and to outperform competitors" (p. 3)

	Strategy Textbook	Definition of Strategy
17.	Johnson, G., Whittington, R. & Scholes, K. (2011) *Exploring strategy*, (9th ed.), Harlow, Essex: Pearson Education.	"Strategy is about key issues for the future of organisations" (p. 3) "Strategy is the long-term direction of an organisation" (p. 3) Strategic management not defined as such as "this book takes the view that managing is always important to strategy. Good strategy is about managing as well as strategising" (p. 7) Strategic planning are the "systematised, step-by-step, procedures to develop an organisation's strategy" (p. 400)
18.	Saloner, G., Shepard, A. & Podolny, J. (2001) *Strategic management*, New York USA: J Wiley & Sons.	"Strategic management…is about developing a set of tools and conceptual maps for uncovering the systematic relationships between the choices the manager makes and the performance the firm realises" (p. 1)
19.	Thompson, A. A., Peteraf, M. A., Gamble, J. E. & Strickland, A. J. (2012) *Creating & executing strategy: The quest for competitive advantage, concepts & cases* (18th ed.), New York: Mc-Graw Hill/Irwin.	"A company's strategy is management's action plan for competing successfully and operating profitably, based on an integrated array of considered choices" (p. 4)
20.	Thompson, J. & Martin, F. (2010) *Strategic management, awareness and change* (6th ed.), Hampshire, UK: South Western Cengage Learning.	"Strategies are means to ends, and these ends concern the purpose and objectives of the organisation. They are the things that businesses do, the paths they follow, and the decisions they take, in order to reach certain points and levels of success." (p. 11)
21.	Viljoen, J. & Dann, S. (2003) *Strategic management* (4th ed.), Frenchs Forest, NSW: Prentice Hall, Pearson Education Australia.	"Strategic management is a term applied to describe those activities of an organisation that enable it to meet the challenges of a constantly changing environment" (p.1)
22.	Wheelan, T. L. & Hunger, D. J. (2010) *Strategic management and business policy: Achieving sustainability* (12th ed.), New Jersey, USA: Prentice Hall.	"A strategy of a corporation forms a comprehensive master plan that states how the corporation will achieve its mission and objectives" (p. 67)

Strategy Textbook	Definition of Strategy
23. Witcher, B. J. & Chau, S. (2010) *Strategic management: Principles and practice* (1st ed.), Mason USA: South Western Cengage Learning.	"an organisation's strategy is an overall approach, or a general pattern of behaviour, for achieving an organisation's purpose, including its strategic objectives" (p. 8)

Content analysis on the language within these definitions enables underlying themes and meanings to be distilled (for example see; Robson, 2002), and reveals a degree of isomorphism regarding definitions of strategy. Thus for an organisation, strategy is about its future, its long-run objectives, its goals and its purpose; further, it is a theory of how those goals, objectives and purposes will be achieved. A definition of strategy that captures the components outlined above is offered by Thompson & Martin (2010), who state that "strategies are means to ends, and these ends concern the purpose and objectives of the organisation. They are the things that businesses do, the paths they follow, and the decisions they take, in order to reach certain points and levels of success" (p. 11).

In attempting to summarise, what is apparent is that strategy is not about what has happened; rather, it is about a future that is not yet realised. In this regard, strategy can be argued as being a future-orientated narrative about how an organisation will achieve its stated purpose. It is a guide to behavior and decision making in the now, but it is also a theory, and in many respects a fiction because it is about the future horizon as much as the unfolding reality.

4. Linking Sustainability and Strategy to Realise a Framework

The earlier discussion on sustainable development revealed it to be a concept regarding humans—our survival, continuation, development and wellbeing. It is a path or purpose for humans to follow, and one that recognises organisations, and in turn corporate strategists, as key actors who set strategic direction. It is future orientated and long term, in so much as it concerns generational timeframes. In this regard, the link between sustainability and understandings of strategy is relatively evident: both concern a narrative about the future and purpose. However, as indicated, in an orgocentric world, where organisations are key actors, the realisation of sustainable development for humanity requires the concept to be embraced by "corporate strategists" (UNSGHLPS, 2012, p. 22), as these individuals "have more opportunity than ever to pick and choose from the best practices and resources... combine them in new and previously unforeseen ways...[and thus help]... to drive sustainable development" (UNSGHLPS, 2012, p. 22). However, an obstacle to such outcomes is determining whether it is corporate strategists' responsibility to pursue such a strategy for humanity, when those individuals may perceive their responsibility as being just their organisation, as a the subset of humanity. Thus, a key concern is how to find a path that enables the strategist to navigate the dichotomy between broad notions of humanity and their organisation as a subset of it.

To navigate this dichotomy, which as will be argued is false, the concept of sustainable development offers a path by calling for a change in humans' understandings. It does this by outlining how sustainable development is not about saving the environment as if the environment is "something outside and completely unrelated to the observer" (Purser Park & Montuori, 1995, p. 1064), and as such an abstract other that exists as a separate concern to human beings. Rather, to realise sustainable outcomes, the concept of sustainability asks us to understand that the world is an "intrinsically dynamic, interconnected web of relations in which there are no absolutely discrete entities and no absolute dividing lines" (Eckersley, 2003, p. 49). Thus, we should not be so naïve to think of the environment as separate to human actions, ambitions and needs (WCED, 1987, p. xi). In other terms, we should not create a false dichotomy between ourselves and the environment. We should recognise the obvious—humans and the rest of nature are mutually embedded. An understanding and acceptance of this is a key inflection point regarding our conceptualisation of the environment, as through it there is no longer a need to perpetuate a Cartesian dualism and a false dichotomy.

In considering the impact on the environment, we should simply consider the impact on ourselves; we should recognise the 'I' in that which we have classically conceived of as merely the environment, and build on Starik and Kanashiro's (2013) argument regarding the nearly obvious that we all breathe, drink and eat the natural environment. In other terms, as Ingold (2011) argues, it surrounds and is entwined with us (Ingold, 2011). Taking this further and to draw out the point more simply, our physical bodies are made of the natural environment. Recognising this entwinement propels us to take concerns about the natural environment as concerns about, for example, what we may eat, breathe and otherwise ingest. While such a move may appear anthropocentric, the challenge is to keep the focus on our physicality. This focus is important, because in literally being made of earth stuff human beings can use their own physical health as a reasonable avatar for other species. For example, if a pollutant in a river is bad for our health, it is likely to be unhealthy for other species that swim in or use the river. If we consider a plant, fish or bird downstream who may use the river, all of those species, like humans, use water to enable their bodily functioning. Similarly they are all composed of organic chemicals, which are combinations of carbon, hydrogen and oxygen among other elements. Consequently, although not likely to always be an entirely accurate avatar, humans can use themselves as a test case for all species and reasonably assume that which does not pollute us as humans is likely not to pollute the fish or bird. In using the individual as an exemplar for the rest of the species on the planet, a dichotomy between the individual and the wider environment is avoided.

To explain further, in pursuing sustainable outcomes a corporate strategist may ask, "What is the impact of an effluent produced by the organisation on a river, or more generally the natural environment?" Such a question ensures that the subject of the question (the river or the natural environment) is an other, a separate conceptual area. Consequently, an ontological dualism is reinforced in how the question is posed. In asking about the impact on the abstract subject, the strategist, an individual who has more capability than many to realise sustainable outcomes, is a separate subject. Further, that individual is abstracted away through the actor being the organisation. Thus the abstract actor is acting on the abstract other, and the decision maker is not within the conceptual domain and either wittingly or unwittingly absolved of responsibility.

The answer to the question depends on trade-offs between what the strategist considers is viable for the organisation and for the abstract other, the environment. The strategist is the judge, but somehow absolved of the act, as if the act that results from the decision is separate to the decision maker, and the actor and acted upon are also separate. In navigating such a trade-off, the strategist is likely to make decisions based upon his perception of what is best for the abstract organisation because the causal link to their own immediate fortunes is more easily conceptualised. Thus notwithstanding that a decision based upon the narrow self-interest of the individual and the organisation is appropriate within current worldviews regarding business.The form of the question propels the individual into a dichotomy that is an ontological fallacy, but real in terms of the decision framework released. Consequently, through the form of the question the individual has to navigate a decision framework where their self-interest is opposed to the other — the river, the environment—as if these domains are human-free zones and not systemically connected.

Thus, while the conception of the environment as a separate concern is analytically useful, it is perpetuating a fractured epistemology that reinforces the separation of humans to everything else. To explain further using an analogous argument, the subject-verb structure of the English language reinforces separation. For example, individuals are likely to say the wind is blowing, as if the wind is a separate action to the blowing. As such a fissure is inserted between the wind and the act of blowing that perpetuates notions of separation as if the wind has attributes that are separate from the blowing. However, the "wind is it's blowing" (Ingold, 2011, p. 17), and similarly the stream is the running water.

To avoid such false fissures being perpetuated, it is necessary to change the form of the questions being asked. Further, in the current epoch of the Anthropocene (Crutzen, 2006), humans are considered the key shapers of the earth. Thus, it is necessary in this epoch to put ourselves in the frame and not abstract ourselves away; it is necessary to put the individual 'I' as the subject given there is no abstract environment, there never was and it was naïve to consider that such a conceptual domain ever existed separately from humans.

As indicated earlier, the use of the 'I' would result in questions that may have typically concerned the natural environment being rephrased to, for instance, 'how will this pollutant impact the air I breathe or the water I drink?' However, this rephrasing then needs to be carried across to the other conceptual domains of society and economy that are typically discussed with regard to sustainable development. Although the use of these labels creates a tension given the arguments above, their use helps ensure analytical comprehensiveness and a systemic perspective. As when previously discussing the environment, the mistake should not be made to treat those domains as abstract, as if they are beyond and devoid of human narrative. When considering these domains, the subject of the question needs to include not just the 'I' but also a personal other, as a society is made up of more than one individual, and hence the individual alone is a limited representation of society relative to environment, and likewise with economic concerns.

When considering societal impacts, concerns regarding wellbeing typically come to the fore. A useful framework for developing the form of questioning is offered by Gladwin, Kennelly and Krause (1995), who argued that sustainable development is a form of human development that is inclusive, connected, equitable, prudent and secure. To précis, Gladwin, Kennelly and Krause (1995) argue that inclusiveness implies time and space, connectivity implies interdependence, equity implies inter-and intragenerational fairness, prudence implies care and prevention, and security implies safety from chronic threats. Consequently, when considering society the questions asked might be: "How does this action by the organisation improve my wellbeing and that of my friends?"; "How does this action enable me to feel more connected to others?"; or, "How does this action enable more opportunities for children in 30 years time?"

Finally ,with regard to economic concerns, the form of questions is similar: "How does this action by the organisation improve monetary outcomes for myself and my neighbours, the people I know?" Table 2 offers a more comprehensive list of questions, though not an exhaustive one. It is important to note that the structure of the questions

I 招待論文/Invited Articles

Table 2　Framework of Questions to Enable Sustainable Outcomes (Questions that do not perpetuate a Cartesian Dualism)

Classic Conceptual Domain	Standard Form of Question– *which would facilitate a dualism between the questioner and the conceptual domain which makes responsibility for the concern relatively easy to absolve because the dualism reinforces an antagonistic relationship*	New Form of Question– *which facilitates monistic, not dualistic understandings and makes responsibility more difficult to absolve because no antagonistic relationship between the questioner and the concern is perpetuated through the form of the question*
Environment	· What is the impact of our products and services on the environment? · What is the impact of our liquid outputs on the environment? · What is the impact of our gaseous outputs on the atmosphere? · Where is the product disposed of? · Will the product decompose through natural processes? · What will be the impact on the local animal population?	· What is the impact of our products and services on us? · How will the liquid outputs impact the water I drink? The sea/river I swim in? · How will the gaseous outputs impact the air I breathe? · Would I want the product disposed of in my neighbourhood? · Would I want the product decomposing in my neighbourhood? Will the decomposition impact the water I drink? · How will the birds I see, the fish I try and catch, the woods I walk in be impacted?...
Society	· How is the product disposed of? · What are the long term effects of this on society? · Will this product or service improve society's wellbeing? · Will this product or service improve society's inclusiveness, security, equity, connectivity?	· Would I be comfortable handling the product in disposable? Would I feel any guilt if my neighbour handled it? · What are the impacts of this strategy on my children? · Will this product or service improve my wellbeing or that of my neighbours, or the people I know in my local vicinity? · Will this product make me or my neighbours more secure, more equitable, more connected, more included?
Economy	· How will this strategy improve the organisation's finances?	· How will this strategy improve my finances and that of the people I buy from? · How can a strategy be developed to enable profitable solutions to those who can't currently afford the products or services?

avoids a fractured epistemology that perpetuates a fractured ontology.

As can be seen, the framework is not comprehensive and requires further development, in that each question leads to supplementary questions that can be added, depending on the answers provided. While every question won't be relevant in any given context, as primers they will encourage an organisation to develop more responsible and sustainable strategies. This is because, as is apparent from the form of the questions, they do not perpetuate separation but rather connection and are, in placing humans and the individual centre stage, thoroughly applicable for our current epoch of the Anthropocene.

5. Summary

This paper has explored a number of areas. First, it outlined the key conceptual arguments regarding sustainable development, exploring how the concept argues for a change in our conceptions away from dualistic to monistic understandings; that is, we should not consider the environment as something separate to human actions and needs. Likewise, we should not consider society and the economy as somehow separate. This discussion also outlined how sustainable development is human-centric and concerns our survival—without our authorship, the concept is pointless.

The second part of the paper outlined understandings of strategy, and how it concerns the long-term purpose and relative success of a focal organisation. The third part of the paper linked these two aspects in an outline for developing sustainable strategies. At the core of this was the centrality of humans to both concepts, and as the ultimate subjects. This centrality was brought forward to highlight how questions about the environment, society and the economy that regard humans as free from interaction are an ontological fallacy. This in turn allowed the development of a new set of questions to drive sustainable strategies in organisations. Thus, rather than asking about an organisation's impact via its products and services on the environment, society or the economy, the strategist needs to ask questions about how the products and services will impact themselves and their associates. This focus on the individual as a subject ensures that a Cartesian dualism is not perpetuated, and enables the strategist to work in recognition and acceptance of systemic interconnections.

The framework in this paper is potentially challenging and confronting for the corporate strategist. However, in the Anthropocene the realisation of sustainable outcomes cannot be outsourced to some benevolent Mother Nature, Mother Society or Mother Economy. Rather, during such a time humans, and the corporate strategists who control the resources to enable sustainable outcomes, have to accept responsibility. This can only begin to happen if the right questions are asked—questions that do not enable individuals to absolve themselves of responsibility but rather to accept it. The form of the question needs to reinforce the questioner's intimate entwinement with all that surrounds them, and thus the potential outcomes of their decisions.

<References>

Banerjee, S. B. (2003) "Who sustains whose development? Sustainable development and the Reinvention of Nature," *Organization Studies*, Vol.24, No.1, pp. 143–180.

Carson, R. (1962) Silent Spring, USA: Houghton Mifflin.

Carruthers, D. (2001) 'From opposition to orthodoxy: The remaking of sustainable development', *Journal of Third World Studies*, Vol.18, No.2, pp. 93–112.

Castree, N. (2002) 'False antitheses? Marxism, nature and actor-networks', *Antipode*, Vol. 34, Issue 1, pp. 111–146.

Costanza, R., Daly, H. E. and Bartholomew, J. A. (1991) 'Goals, agenda, and policy recommendations for ecological economics' *Ecological economics: the science and management of sustainability*, New York, NY: Columbia University Press. pp. 1–20, fcostanza.

Crutzen, P. (2006) The 'Anthropocene', in E. Ehlers & T. Krafft (Eds.), *Earth system science in the Anthropocene*, Heidelberg, Berlin: Springer.

Eckersley, R. (2003) *Environmentalism and Political Theory: Towards an Ecocentric Approach*, London: Routledge. (reprint of 1992 edition)

Egri, C. P. & Pinfield, L. T. (1999) 'Organizations and the biosphere: Ecologies and environments' in S. R. Clegg', C. Hardy & W. R. Nord (Eds.), *Managing Organizations*, pp. 209–233. London: Sage.

Ehrlich, P. (1971) *The Population Bomb*, London: Pan.

Fiol, C. M. (1989) 'A Semiotic analysis of corporate language: Organizational Boundaries and Joint

Venturing', *Administrative Science Quarterly*, Vol.34, No.2, pp. 277-303.

Gladwin, T. N., Kennelly, J. J. and Krause, T. S. (1995) 'Shifting paradigms for sustainable development: Implications for management theory and research', *Academy of Management Review*, Vol. 20, No.4, pp. 874-907.

Guattari, F. (1989) 'The three ecologies', *New Formations*, Issue 8 (summer), pp. 131-147.

Hardin, G. (1974) 'Lifeboat Ethics: the case against helping the poor', *Psychology Today*, http://www.garretthardinsociety.org/articles/art_lifeboat_ethics_case_against_helping_poor.html, Accessed 14th November, 2014.

Hopkins, M. S. (2009) '8 Reasons Sustainability will Change Management (That You Never Thought of)', MIT *Sloan Management Review*, Vol.51, Issue1, pp. 27-30.

Ingold, T. (2011) *Being alive: Essays on movement, knowledge and description*, Abingdon, UK: Routledge.

Kiron, D., Kruschwitz, N., Haanaes, K. and von Streng Velken, I. (2012) 'Sustainability Nears a Tipping Point', *MIT Sloan Management Review*, Vol.53, Issue 2, pp. 69-74.

———, ———, Reeves, M. and Goh, E. (2013) 'The Benefits of Sustainability-Driven Innovation', *MIT Sloan Management Review*, Vol. 54, Issue 2, pp. 69-73.

Lakoff, G. (2010) 'Why it Matters How We Frame the Environment', *Environmental Communication*, Vol.4, Issue 1, pp. 70-81.

Latour, B. (1999a) 'On recalling ANT', In J. Law & J. Hassard (eds.), Actor *Network Theory and After*, Oxford: Blackwell.

——— (1999b) Pandora's *hope essays on the reality of science studies*, London: Harvard University Press.

March, J. G. & Simon, H. A. (1958) *Organizations*, New York: Wiley.

Meadows, D. H., Meadows, D. L., Randers, J., Behrens III, W. and Visser, W. (1972) *The Limits to Growth: a report to the Club of Rome's project on the predicament of mankind*, New York: Universe.

Newton, T. (2002) 'Creating new ecological order? Elias and actor-network theory', *Academy of Management Review*, Vol.27, No.4, pp. 523-540.

Purser, R. E., Park, C. and Montuori, A. (1995) 'Limits to anthropocentrism: Toward an ecocentric organization paradigm?', *Academy of Management Review*, Vol.20, No.4, pp. 1053-1089.

Robson, C. (2002) *Real world research* (2nd ed.), Oxford: Blackwell Publishing.

Shrivastava, P. and Hart, G. (1994) 'Greening organizations—2000', *International Journal of Public Administration*, Vol.17, Issue 3-4, pp. 607-635.

Simon, H. A. (1947) *Administrative behavior: A study of decision-making processes in administrative organizations*, New York: Simon & Schuster.

Starik, M. and Kanashiro, P. (2013) 'Toward a Theory of Sustainability Management: Uncovering and

Integrating the Nearly Obvious', *Organization & Environment*, Vol.26, No.1, pp. 7-30.

Steer, A. and Wade-Gery, W. (1993) 'Sustainable development: Theory and practice for a sustainable future', *Sustainable Development*, Vol.1, No.3, pp. 23-35.

Spender, J. C. (1989) *Industry recipes: The nature and sources of managerial judgment*, Oxford: Basil Blackwell.

Suzuki, D. (2008) *The Sacred Balance: rediscovering our place in nature*, Sydney: Allen & Unwin.

Tulloch, L. (2013) 'On Science, Ecology and Environmentalism.', *Policy Futures in Education*, Vol.11, No.1, pp. 100-114.

―― and Neilson, D. (2014) 'The Neoliberalisation of Sustainability', *Citizenship, Social and Economics Education*, Vol.13, No.1, pp. 26-38.

United Nations Secretary—General's High Level Panel on Global Sustainability (2012) *Resilient people, resilient planet: A future worth choosing*, New York: United Nations.

Whittington, R. (1993) *What is strategy—and does it matter?*, London: Routledge.

World Commission on Environment and Development (1987) *Our common future*, Oxford: Oxford University Press.

Yates, J. J. (2012) 'Abundance on trial: The cultural significance of sustainability', *The Hedgehog Review*, Summer, pp. 8-25.

3 Corporate Greening
―― A Conceptual Analysis

Gabriel Eweje
Associate Professor, School of Management, Massey Business School, Massey University,
New Zealand

Aymen Sajjad
PhD candidate, School of Management, Massey Business School, Massey University,
New Zealand

Mina Sakaki
Citizen Participation Promotion Division, Domestic Strategy and Partnership Department,
Japan International Cooperation Agency

[Abstract]
Increasingly, there has been a great deal of attention paid to the notion of corporate sustainability which has variously been defined as the incorporation of social, environmental, economic and cultural concerns into corporate strategy. As such, business participation in social and environmental responsibility has moved from being presented as an opportunity to 'give back' something to society to action that can enhance business performance. Accordingly, many businesses have increased their sustainability initiatives in order to prevent a backlash from, and counter negative perceptions held by, various stakeholders about business activities. This shifts responsibility from being something that is a nice thing to do to something that can be justified on the basis of economic self-interest. Demonstrating that a business case exists for responsibility counters the objection to businesses moving outside the traditional areas of activity that have been held to impact on shareholder value. Thus, this conceptual chapter discusses the greening of business which we describe as a set of initiatives that focus on mitigating company's social and environmental impacts as well as improving its competitive advantage.

Keywords : green management, corporate greening, stakeholder, green design, green operations, life cycle management, cleaner production, green operations, green supply chain management, integrative greening strategies.

1. Introduction

Sustainability is no more a new concept in management, however, it is still an evolving corporate management paradigm, and there is an increasing recognition that organisations must address the issue of sustainability in their operations (Ahi and Searcy, 2013). Although the concept acknowledges the need for profitability, it differs from the traditional growth and profit-maximisation model in that it places a much greater emphasis on environmental, social, and economic performance, and the public reporting on this performance. In addition, companies in the 21st Century are struggling with a rapidly changing world where more consumers and other stakeholders are increasingly aware of the sustainability notion. Thus, a great deal of attention has been paid to the notion of sustainability, which has variously been defined as the incorporation of social, environmental, economic and cultural concerns into corporate strategy. As such, business participation in social and environmental responsibility has moved from being presented as an opportunity to 'give back' something to society to action that can enhance business performance. Accordingly, the emphasis has shifted to companies to be more proactive in dealing with both environmental and social issues in order to remain competitive. As argued by Hart (1997, p. 67) "environmental sustainability—the need to protect the environment and conserve natural resources—is now a value embraced by the most competitive and successful multinational".

This chapter discusses briefly the conceptualisation of green management, the role of business in mitigating environmental sustainability impacts, and initiatives and strategies put in place to reduce companies' environmental impacts.

2. What is Corporate Greening?

Over recent years, the notion of 'environmental responsibility' has expanded beyond compliance with regulations and initiatives like recycling and energy efficiency (Eweje, 2006). He further posited that environmental organisations, company managers and consumers now view environmental responsibility as involving a comprehensive approach that includes assessing business products, eliminating waste and emissions; maximising efficiency and avoiding practices that damage the environment.

It is apposite at this juncture to briefly describe the notion of green management. The need for environmental awareness and green management evolves from a variety of wrongdoings that have occurred over time. Academics and practitioners' perspectives of corporate greening ranges from "simple and basic environmentally-friendly programs that prevent further harm, to complex and demanding strategic initiatives that help to restore the environmental damage that has been done in the past" (Haden, Oyler and Humphreys, 2009, p. 1042). As a result, businesses are increasingly being required to align with societal norms while generating financial returns (Ditlev-Simonsen and Midttun, 2011; Parmar et al., 2010; Steurer, Langer, Konrad, and Martinuzzi, 2005; Wolf, 2014).

Consequently, in this chapter, we go a step further by incorporating social issues into our discussion of green management. For example, social sustainability refers to as sustaining the well-being of people, while environmental sustainability refers to sustaining nature or natural resources (Kopnina and Blewitt, 2015). They argue that the two concepts are intricately interlinked as human welfare depends on the sustainability of the environment. Thus, we describe green management as 'a set of initiatives that focus on mitigating company's social and environmental impacts with deeper understanding of our natural environment and the role of business within it'. This view resonances with Haden et al. (2009, p. 1052) who define green management as the "organization-wide process of applying innovation to achieve sustainability, waste reduction, social responsibility, and a competitive advantage via continuous learning and development and by embracing environmental goals and strategies that are fully integrated with the goals and strategies of the organization." Other scholars such as Peng and Lin (2008, p. 203) define green management as "practices that produce environmentally-friendly products and minimize the impact on the environment through green production, green research and development, and green marketing". Following this argument, Banerjee (1998) proposes that organizations can gain a competitive advantage by learning how to integrate environmental issues with their business strategies and corporate goals.

One of the challenges facing corporations is how to define their relationship to the natural environment (Larson, 2000). According to this viewpoint, it could be argued that "management efforts that characterised the 1970s through the 1990s have led to the emergence of the concept of corporate environmental responsibility within the sustainability framework" (Larson, 2000, p. 304). Hence, the pressing environmental issues have prompted responses from different quarters such as business practitioners, international organisations, non-governmental organisations (NGOs), and academics that we have to change the patterns of production and consumption. Environmental problems range from the processes of industrialization, to population growth, and waste management. Accordingly, environmental issues include climate change, depletion of natural resources, biodiversity loss and pollution (Esty and Simmons, 2011; Esty and Winston, 2009). All the aforementioned issues directly affect our environment and well-being and as such businesses have been called upon to mitigate the problems. Therefore, while governments' intervention can be good to exert pressure on businesses, it is argued in this chapter that business efforts and initiatives are necessary and fundamental in reducing environmental impacts in the long term. There is no doubt that the need for governments' intervention in terms of regulation gets widespread but grudging acceptance, however, it has been suggested that environmental regulations erode competiveness (Porter and van der Linde, 1995).

Dobers and Wolff (2000) have listed regulation, consumer awareness, companies' solution of end-of-pipe problems, and companies' green product development as the most important element in achieving sustainability or rather, green management. They argue that regulation has been directed to control the physical resources in production and the consumption process. The argument is that environmental regulation should control business production to minimize pollution and resource use, and that consumers should be educated to consume environmentally friendly products. Further, regulation and consumer awareness should influence what companies do; companies then would influence the pull in the market by providing adequate products. Dobers and Wolff (2000) further assert that environmental awareness in society has indicated many activists and companies are of the opinion that consumers behave according to their expressed values, which would simply imply that they consume better environmentally adapted products.

In addition, voluntary environmental reporting has been encouraged for transparency (Clarkson, Overell and Chapple, 2011; Kolk, 2005). The idea is to make business more transparent for companies' stakeholders such as customers, NGOs, governments, employees, and suppliers as well as the financial market. It is argued that the fundamental idea of the financial market inclusion is that financial institutions should screen industry from an environmental point of view to judge and evaluate environmental risks and price companies more on their actual liabilities.

Berry and Rondinelli (1998) have argued that progressive companies are moving away from a strategy of regulatory compliance to one of proactive management in order to remain competitive. They argued: "environmental responsibility—the need to protect the environment and conserve natural resources—is now a value embraced by the most competitive and successful multinational" (p. 38). Non-compliance can be costly, and lead to legal and ethical crises that could be more expensive and loss of profits. However, many companies are investing on social and environmental projects to stay abreast of environmental regulations. Corporations are looking beyond complying to enhance ethical images, avoid serious legal liabilities, satisfy the safety concerns of employees, respond to government regulations and shareholders, and develop new business opportunities in order to remain competitive in world markets (Berry and Rondinelli, 1998), as well as producing goods and services under acceptable environmental conditions, and refrain from causing social and environmental problems.

Further on government environmental regulations, Porter and van der Linde (1995), have argued that the static view of environmental regulation in which everything except regulation is held constant is incorrect. They argue that if technology, products, processes, and customer needs were all fixed, the conclusion that regulation must raise costs would be inevitable. They assert that since companies operate in a real world of dynamic competition, corporations are "constantly finding innovative solutions to pressures of all sorts—from competitors, customers, and regulators" (Porter and van der Linde, 1995, p. 120). It was further suggested that properly designed environmental standards can trigger innovations that lower the total costs of a product or improve its value. Accordingly, such innovations allow companies to use a range of inputs more productively—from raw materials to energy to labour thereby offsetting the costs of improving environmental impact and ending the stalemate. Thus, this "enhanced resource productivity makes companies more competitive, not less" (Porter and van der Linde, 1995, p. 121).

Proactive environmental strategies such as reducing waste, natural resources, and cutting costs can simultaneously respond to customer and shareholder demands. Thus, companies seeking to satisfy diverse stakeholders have realised that proactive environmental management demands more than adjustments to government policies. The strategies may require firms to make more effective use of corporate intelligence to define more missions, realign company value systems, find new approach to managing change, accelerate training and education, and modify behaviour throughout the organisation (Berry and Rondinelli, 1998). They further argue that for many firms the challenge is to balance concerns with cash flow, profitability, and environmental protection in order to respond to the demands of increasingly diverse groups of stakeholders. Accordingly, "many companies that adopted quality management programs to improve competitive positions—3M, Kodak, Sony, Alcoa, Volvo, Procter and Gamble—are recognised by their stakeholders for exemplary environmental performance" (Berry and

Rondinelli, 1998, p. 41). Hence, we argue in this chapter that adopting proactive environmental management practices and voluntary standards, and integrating such initiatives to corporate strategy will enhance competitiveness.

3. Corporate Greening Strategies

Scholars have proposed a range of green strategies and principles to enhance corporate environmental performance. As stated earlier, the implementation of green strategies in modern corporations is predominantly triggered by stakeholders' activism towards harmful environmental impacts of production activities (Gupta, 1995; Hughes, Anderson and Golden, 2001; Rusinko, 2007). Furthermore, several recent studies have reported that there exist a positive association between corporate green investments and competitive advantage (Molina-Azorin, Claver-Corte´s, Pereira-Moliner, and Tarı´, 2009; Rao and Holt, 2005). Klassen and Vachon (2003) assert that companies can reduce their environmental impacts by increasing investments in green technologies, and diverting investments from pollution control to pollution prevention environmental methods. Thus, such advanced pollution prevention technologies and proactive management of environmental impacts can reduce waste, redundancies, inefficiencies, costs, natural resources usage, which will lead to improve the business competitiveness.

Companies can employ a number of green strategies to mitigate their negative environmental impacts and improve their environmental performance. However, in this chapter these strategies are clustered into three main types under a broad theme refers to as 'the green management'. These strategies include green-design and life cycle management (LCM) (Eltayeb, Zailani and Ramayah, 2011; Hervani, Helms and Sarkis, 2005), green operations and cleaner production (Fresner, 1998; Rao, 2004; Srivastava, 2007), and green supply chain management (GSCM) (Green Jr, Zelbst, Meacham and Bhadauria, 2012; Holt and Ghobadian, 2009; Rao and Holt, 2005). The subsequent discussion critically examines these bundles of green strategies in detail. In particular, the focus of the discussion is to explore how companies can manage their environmental impacts and enhance green performance. Notwithstanding, it is important to note that different sectors have different requirements and thus implementation of these strategies will vary from sector to sector depending on contextual variables such as management approach, company size, management systems, regulatory regimes, market factors, industry dynamics or other business requirements.

3-1. Green-design, Life Cycle Management and Product Stewardship

Green-design or design for the environment (DFE) relates to the adoption of environmentally conscious practices in the product development process by which harmful environmental impacts of a product can be mitigated or totally eliminated during its total life cycle (Ashby, Leat and Hudson-Smith, 2012; Gertsakis, Ryan and Lewis, 1997; Hervani et al., 2005). Eltayeb, Zailani and Ramayah (2011, p. 497) defined green design as "actions taken during product development aim at minimizing a product's environmental impact during its whole life cycle—from acquiring materials, to manufacturing, use, and ultimately to its final disposal".

The scope of green design strategy may transcend into different disciplines such as engineering, supply chain management, occupational health and safety and environmental management (Soylu and Dumville, 2011; Srivastava, 2007). The management of cross-disciplinary integration among these disciplines is an important aspect for the implementation of green design perspective. Close co-ordination among people working in these disciplines or working areas is a prerequisite in terms of reaping environmental advantages associated with green product. In fact, transforming idea of green product from 'conceptualization into reality' requires integrated effort and collaboration among various departments within a company and partners across supply chain network.

It is pertinent to note that green-design activities are focused on both the development of green products and green processes (Tsoulfas and Pappis, 2006). Green design principle provides an opportunity for proactive companies to conceptualize, design and develop innovative products which are ecologically harmonious, recoverable, durable, recyclable or decomposable (Tsoulfas and Pappis, 2006). For example, green design strategies including design for recycling, design for longer life, design for resource conservation, design for disassembly and design for cleaner manufacturing and design for pollution prevention are instrumental in designing products in a way that they can be easily reused, recycled and remanufactured at the end-of-product life (Esty and Simmons, 2011; Gertsakis et al., 1997; Kopnina and Blewitt, 2015).

Sarkis (2003) argues that the introductory phase of product development has a critical role in the development of green products. Embedding green thinking into a new product development process, and taking necessary actions at the beginning of the development of new products or processes would significantly reduce environmental impacts of a product during its life cycle phases (Handfield, Melnyk, Calantone and Curkovic, 2001). Therefore, an integration of green thinking would not only enable companies to manufacture green products, achieve costs savings, operational efficiencies, and reduce their environmental impacts (Esty and Simmons, 2011) but at the same time stimulate responsible product use and disposal from consumers' perspective.

LCM is an essential complementary framework to green design strategy (Donnelly, Beckett-Furnell, Traeger, Okrasinski and Holman, 2006), and contributes to companies' efforts in achieving a more sustainable product development goal (Saur et al., 2003). It is also referred to as 'cradle-to-grave' method which examines the industrial systems involved in the entire life cycle of a product. Furthermore, LCM framework enables companies to understand the long-term social, environmental and economic impacts that relate to the production of their products. Thus, better understanding of their impact would assist practitioners to make informed production decisions that result in improved sustainability performance and risk management (Epstein, 2008). Similarly, Saur et al. (2003) assert that LCM is the integration of life cycle thinking into business strategy, practices, and decision-making processes with the aim to rationalize and manage products and services life cycle impacts in a way that promote sustainable consumption and production choices.

According to Hervani et al. (2005, p. 336), "LCM is a structural approach to define and evaluate the total environmental load associated with providing a service. It also incorporates development of an inventory of data, impact of materials, products and processes, and improvement analysis aspects". In a same vein, the United States Environmental Protection Agency (2014) acknowledges that LCM involves an organization's concern for the entire product life cycle and consideration is given to various stages of a product's supply chain including raw material extinction, logistics and transportation aspects, cleaner production, waste management, product repair and use as well as its end-of-life management. Thus, inputs, outputs and potential environmental impacts of a particular product are systematically complied and examined, which lead to informed decision making process and progress towards desired sustainability outcomes (Donnelly et al., 2006).

In addition, some scholars have argued in support of more sophisticated green design approaches such as product stewardship that emphasized on cradle to cradle (regenerative thinking) paradigm (McDonough and Braungart, 2010) and an industrial ecology principle (Allenby, 1999; Frosch and Gallopoulos, 1989; Seuring, 2004). According to McDonough et al. (2003), cradle to cradle paradigm requires industrial systems to be commercially productive, socially beneficial, and ecologically intelligent. The idea is to develop products in such a way that industrial systems optimize energy and material use and reduce the amount of waste and pollution at their lowest level (Frosch and Gallopoulos, 1989) by utilizing each other's material and by-products so that nothing get wasted in the system. Consequently, the notion of product stewardship substantiates an optimal utilization of resources and circular production model in which natural resources circulate in a system without any harmful impacts on the eco-system and the natural environment. In other words, industrial ecology and product stewardship concepts advocate zero waste principle in which "waste from one industrial process can serve as the raw materials for another, thereby reducing the impact of industry on the environment" (Frosch and Gallopoulos, 1989, p. 1).

3-2. Green Operations and Cleaner Production

The concepts of green operations and cleaner production refer to the application of innovative production methods that deliver improved economic, social and environmental performance while producing goods and services for customers. As a rule of thumb, those companies which are beginners in the adoption of sustainability principles can identify and attain immediate advantages from low hanging fruits (Esty and Simmons, 2011) by energy savings and elimination of redundancies in the production of good and services. Therefore, embracing lean manufacturing strategy would be a good starting point for such companies. The lean manufacturing is defined as "a set of techniques that aim to eliminate each form of waste along the value chain. These techniques are clustered into bundles of practices, such as just in time (JIT), total quality management (TQM), and total preventive maintenance (TPM), which, on the whole, implement the lean philosophy of waste elimination and continuous improvement (Galeazzo, Furlan and Vinelli, 2014, p. 192). These practices are primarily aimed at improving the economic and operational performance of companies by curtailing non-value adding activities in the production process (Corbett and Klassen, 2006; Rao, 2004), however, some environmental benefits may also be accompanied by the implementation of lean manufacturing practices (Kleindorfer, Singhal and Wassenhove, 2005). Accordingly, certain lean manufacturing practices can be complementary to green management and synergetic linkages can be established to improve environmental performance (Dües, Tan and Lim, 2013).

Furthermore, green manufacturing practices consist of those set of production methods which enable companies to reduce, mitigate or limit the negative environmental impacts. In addition, such methods enable the reduction of natural resources usage and promote sustainable consumption of products and services. According to Srivastava (2007, p. 55), "green operations relate to all aspects related to product manufacture/re-manufacture, usage, handling, logistics and waste management once the design has been finalized". The key green manufacturing and operations management strategies include eco-efficiency and cleaner production.

The concept of eco-efficiency focuses on 'doing more with less'. Some scholars view eco-efficiency as proactive business approach that links environmental excellence with business excellence (DeSimone and Popoff, 2000). Contrary to the traditional production management perspective, eco-efficiency stimulates a greater value in production systems by integrating simple but useful principles such as reduce, reuse and recycle (DeSimone and Popoff, 2000). The world business council for sustainable development (WBCSD) defined the concept of eco-efficiency as "the delivery of competitively priced goods and services that satisfy human needs and bring quality of life, while progressively reducing ecological impacts and resource intensity throughout the life-cycle to a level at least in line with Earth's estimated carrying capacity" (WBCSD, 1996, p. 5). Thus, the primary goal of this approach is to create products or services in a way that optimizes resource utilization through innovative production methods, while at the same time waste, pollution and redundancies can be eliminated or minimized to its lowest possible levels (Rao, 2004).

Another comparable strategy to eco-efficiency is cleaner production which is defined as "a systematically organized approach to production activities, which has positive effects on the environment. These activities encompass resource use minimization, improved eco-efficiency and source reduction, in order to improve the environmental protection and to reduce risks to living organisms" (Glavič and Lukman, 2007, p. 1879). According to Glavič and Lukman (2007), this definition can be applied to cleaner production processes, services and green products. Thus, cleaner production approach is an integrated preventive environmental strategy which does not only enhances eco-efficiency in the production processes but also minimizes the environmental risks to society and the natural environment. The WBCSD (1996) maintains that implementation of cleaner production strategy requires change in attitudes, responsible environmental management and continuous search for new creative technologies that enhance environmental performance. There exist strong interlinkages between eco-efficiency and cleaner production concepts. For instance, both of these concepts aim at concurrently developing business excellence and environmental excellence.

3 Corporate Greening 107

The application of environmental management systems (EMSs) is an important step in the implementation of eco-efficiency and cleaner production strategies in companies. Past studies suggest that implementing EMSs in organizations can lead to several advantages including improved environmental performance, corporate image, operational efficiency, risk minimization, better legal compliance, minimization of future liabilities, and most importantly a 'win-win' potential for economic and ecological benefits (Melnyk, Sroufe and Calantone, 2003; Potoski and Prakash, 2005; Stapleton, Glover and Davis, 2001; Steger, 2000). An EMS is defined as "a transparent, systematic process known corporate-wide, with the purpose of prescribing and implementing environmental goals, policies, and responsibilities, as well as regular auditing of its elements" (Steger, 2000, p. 24). Increasingly, domestic and multinational corporations (MNCs) around the world are either developing their own EMSs or adopting international environmental certifications because these standards are more or less becoming an essential selling point in the global market. Especially, a large number of MNCs demand these environmental certifications from supplying companies to demonstrate their environmental commitment and reduce business risk (Morrow and Rondinelli, 2002).

Accordingly, once a company implements EMS then it may further go down the route of certified EMS standard (Darnall, Jason Jolley and Handfield, 2008). The two most widely used and internationally accepted EMS standards are International Organization for Standardization (ISO) 14001 and eco-management audit scheme (EMAS) (Morrow and Rondinelli, 2002). The ISO 14001 is an international certification developed in 1996 and for a company to become ISO 14001 certified, it has to go through an independent third party audit (TPA) mechanism (Darnall et al., 2008). The purpose of the TPA system is to guarantee stakeholders that a certified ISO 14001 organization conforms to relevant standards. However, it is important to note that ISO 14001 is not a performance oriented standard. This means that ISO 14001 certified company does not require a specific level of environmental performance.

On the other hand, EMAS is the European Union's voluntary standard which is developed in 1993 (Darnall et al., 2008; Morrow and Rondinelli, 2002). The purpose of the

EMAS to promote continuous environmental improvement, environmental efficiency, and publish their environmental achievements (Glavič and Lukman, 2007). EMAS was originally designed for manufacturing organizations, however, in 2001 the European parliament recommended that EMAS can be applicable to all organizations, both public and private sectors, which have environmental impacts (Glavič and Lukman, 2007). The certified EMAS organizations need to maintain and improve environmental performance on a continual basis. In addition, the majority of companies holding EMAS certification are in Europe and yet this certification has not achieved a similar recognition as ISO 14001. However, it is important to recognize that EMSs focus on internal organizational environmental performance but the external environmental management aspects such as suppliers' environmental and regulatory compliance issues, logistics and distribution management activities, product end-of-life management and many other environmental supply chain management (SCM) facets are not addressed through EMS's implementation (Darnall et al., 2008). Thus, the following section examines the key GSCM practices and strategies which companies can use to enhance their SCM performance.

3-3. Green Supply Chain Management

Green Supply chain management (GSCM) is relatively a broad notion that relates to greening the entire supply chain activities of a company. It acknowledges the significance of developing environmental performance of all supply chain actors involved in the production of good and services rather than only concentrating on the production processes. In other words, GSCM perspective encourages an integrated environmental thinking and joint environmental responsibility among the key actors involved in the entire value chain that delivers an end product to consumers as well as its reverse product flows (Corbett and Klassen, 2006). Srivastava (2007, pp. 54-55) defined GSCM as "integrating environmental thinking into supply-chain management, including product design, material sourcing and selection, manufacturing processes, delivery of the final product to the consumers as well as end-of-life management of the product after its useful life". In the previous discussion two aspects of the GSCM, that is, green design, and green operations management have been discussed in detail. Therefore, the subsequent discussion critically examines other key strategies including green procurement, green logistics management, and close loop supply chain management in detail.

Traditionally, the role of procurement function was to procure those materials, products or services, which satisfy commercial standards such as cost, quality and on-time delivery. However, the notion of green procurement emphasises on embedding environmental sustainability criteria along with traditional standards into organizational purchase decisions (Eltayeb et al., 2011). It is recognized that procurement is a vital boundary-spanning function that has an important role in the development of SCM environmental capability by reducing emissions and pollution (Blome, Hollos and Paulraj, 2014). Scholars have provided different definitions and different terminologies have been used to conceptualize the concept of green procurement such as green supply management, green purchasing and environmental purchasing. Zsidisin and Siferd (2001, p.69) define "environmental purchasing for an individual firm is the set of purchasing policies held, actions taken, and relationships formed in response to concerns associated with the natural environment. These concerns relate to the acquisition of raw materials, including supplier selection, evaluation and development; suppliers' operations; inbound distribution; packaging; recycling; reuse; resource reduction; and final disposal of the firm's products". Zsidisin and Siferd (2001) further suggest that purchasing practitioners are in an outstanding position to influence the concerns about the natural environment. Essentially, by placing a robust environmental screening criteria and preliminary environmental control systems, purchase management function can potentially limit the future environmental harms and other undesirable incidences in other SCM.

Green logistics is a key area that plays a vital role in achieving GSCM. Greening of logistics activities encompasses safeguarding and minimizing the environmental impacts such as GHGs emission, pollution, fuel usage, and other forms of waste that are harmful to the natural environmental and human health (Grant, Trautrims and Wong,

2013). Formerly, logistics function was perceived as one-way directional processes involving moving a good or raw materials from the point of extraction to manufacturing and to the end consumer. The main goals of logistics function were to reduce costs, improved service delivery and customer satisfaction. Therefore, issues such as greening the logistics and reverse logistics flows relating to product end-of-life were not considered as significant concerns for logistics professionals in the past. However, contemporary growing stakeholders concerns around environmental issues such as food miles debate, transport impacts, and carbon footprints have made green logistics an important area to logistics, sustainability and SCM practitioners. Accordingly, a range of different logistics approaches have been proposed to manage negative environmental impacts of logistics activities. Some of the key logistics strategies include green forward logistics and reverse logistics, close loop supply chain management (CLSCM),

Improved environmental performance of logistics activities depends on both forward and reverse logistics flows. The forward flows relate to taking raw materials from the point of origin to end use of the product excluding product end-of-life management. A range of techniques can be utilized to reduce environmental impacts emerging from forward logistics flows. Some of the useful green logistics methods to enhance green management include slow streaming, alternative transport modes, green warehousing, smart green packaging design, technological advancement (e.g., use of GPS and automatic engine shut down devices) and application of ICT in logistics management. Others are network redesign for sustainability, collaboration with logistics service providers, consolidation of freight and route sharing, and the use of alternative sustainable fuels options (Dauvergne and Lister, 2013; Dey, LaGuardia and Srinivasan, 2011; Grant et al., 2013; Oberhofer and Dieplinger, 2014). All these methods hold a great potential to make logistics activity environmentally and economically sustainable.

On the other hand, reverse flows relate to end-of-life management of a product involving activities such as recycling, resale, remanufacturing, refurbishing, reuse efforts in the organizational SCM. Rogers and Tibben-Lembke (1999, p.2) define reverse logistics as "the process of planning, implementing, and controlling the efficient, cost effective flow of raw materials, in-process inventory, finished goods, and related information from the point of consumption to the point of origin for the purpose of recapturing or creating value or proper disposal". According to Sarkis, Helms and Hervani (2010), reverse logistics is an important environmental strategy that enable companies to decelerate or prevent environmental degradation. In addition, Sarkis (2003) argues that an appropriate distribution and logistics decisions such as choice of distribution outlet location, transport modes and logistics control mechanisms can impact both reverse and forward logistics networks in terms of environmental sustainability. Therefore, the purpose of synchronized application of green forward and reverse logistics flows is to achieve CLSCM (Nikolaou, Evangelinos, and Allan, 2013) where the circular production thinking can be infused across the SCM network as well as developing economic and environmental performance of the businesses that network (Carter and Ellram, 1998; Nikolaou et al., 2013). CLSCM concept relates to the design, control, and operation of a system to maximize value creation over the entire life cycle of a product with dynamic recovery of value from different types and volumes of returns over time" (Guide Jr and Van Wassenhove, 2009, p. 10).

It is fundamental that companies integrate all the aforementioned strategies holistically (see Fig.1) and not as a unit or individually in order to achieve competitive advan-

Fig. 1 Integrative Greening Strategies

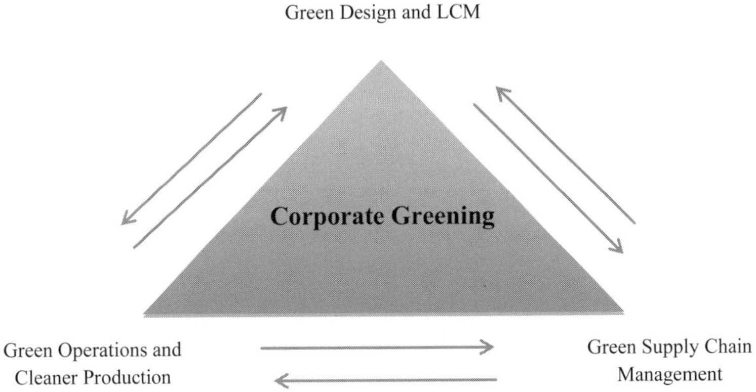

Green Design and LCM

Corporate Greening

Green Operations and Cleaner Production

Green Supply Chain Management

tage with the greening strategies discussed. Focusing on one strategy at a time will not achieve the same benefit as integrating all the strategies into business activities at once. Thus, the holistic view of the strategies discussed above is particularly encouraged.

4. Conclusion

This chapter has discussed the role of green management in the modern business environment. It is argued that business participation in social and environmental responsibility has moved from being presented as an opportunity to 'give back' something to society to action that can enhance business performance. Integrating environmental thinking into business strategies and decision making processes is imperative in order to sustain the long-term survival of a company in the marketplace. Furthermore, an effort has been made to demonstrate that a business case exists for responsibility. This counters the objection to businesses moving outside the traditional areas of activity that have been held to impact on shareholders' value. Investment in social and environmental responsibility can simultaneously respond to stakeholder concerns while generating a positive financial return for shareholders.

Building on corporate greening paradigm, three key green strategies—green-design and life cycle management (LCM), green operations and cleaner production, and GSCM—are discussed in this chapter. It is argued that corporate greening efforts must be holistic taking into account various stages of production activities. Each one of these strategies plays an important role in developing a company's environmental performance. However, simultaneous improvements in environmental, social and economic performance are contingent on how well a company integrate these strategies while performing business activities.

<References>

Ahi, P. and Searcy, C. (2013) 'A comparative literature analysis of definitions for green and sustainable supply chain management', *Journal of Cleaner Production*, Vol.52, pp. 329–341.

Allenby, Braden. R. (1999) *Industrial Ecology: Policy and Implementation*, NJ: Prentice-Hal.

Ashby, A., Leat, M. and Hudson-Smith, M. (2012) 'Making connections: a review of supply chain management and sustainability literature', *Supply Chain Management: An International Journal*, Vol.17, Issue 5, pp. 497–516.

Banerjee, S. B. (1998) 'Corporate environmentalism perspectives from organizational learning', *Management Learning*, Vol.29, No.2, pp. 147–164.

Berry, M. A. and Rondinelli, D. A. (1998) 'Procative corporate enviornmental management: a new industrial revolution', *Academy of Management Executive*, Vol.12, No.2, pp. 38–50.

Blome, C., Hollos, D. and Paulraj, A. (2014) 'Green procurement and green supplier development: antecedents and effects on supplier performance', *International Journal of Production Research*, Vol.52, No.1, 32–49.

Carter, C. R. and Ellram, L. M. (1998) ' Reverse logictics-A reveiw of the literature and framework for future investigation', *Journal of Business Logistics*, Vol.19, pp. 85–102.

Clarkson, P. M., Overell, M. B. and Chapple, L. (2011) 'Environmental reporting and its relation to corporate environmental performance', *Abacus*, Vol.47, No.1, pp. 27–60.

Corbett, C. J. and Klassen, R. D. (2006) 'Extending the horizons: environmental excellence as key to improving operations', *Manufacturing and Service Operations Management*, Vol.8, No.1, pp. 5–22.

Darnall, N., Jason Jolley, G. and Handfield, R. (2008) 'Environmental management systems and green supply chain management: complements for sustainability?', *Business Strategy and the Environment*, Vol.18, pp.30–45.

Dauvergne, P. and Lister, J. (2013) *Eco-Business: a big-brand takeover of sustainability*, Cambridge, MA: MIT Press.

DeSimone, L. D. and Popoff, F. (2000) *Eco-efficiency: the business link to sustainable development*, Cambridge, MA: MIT press.

Dey, A., LaGuardia, P. and Srinivasan, M. (2011) 'Building sustainability in logistics operations: a research agenda', *Management Research Review*, Vol.34, No.11, pp. 1237–1259.

Ditlev-Simonsen, C. D. and Midttun, A. (2011) 'What motivates managers to pursue corporate responsibility? A survey among key stakeholders', *Corporate Social Responsibility and Environmental Management*, Vol.18, No.1, pp. 25–38.

Dobers, P. and Wolff, R. (2000) 'Competing with 'soft'issues—from managing the environment to sustainable business strategies', *Business Strategy and the Environment*, Vol.9, No. 3, pp. 143–150.

Donnelly, K., Beckett-Furnell, Z., Traeger, S., Okrasinski, T. and Holman, S. (2006) 'Eco-design implemented through a product-based environmental management system', *Journal of Cleaner Production*, Vol.14, No.15, pp. 1357-1367.

Dües, C. M., Tan, K. H. and Lim, M. (2013) 'Green as the new Lean: how to use Lean practices as a catalyst to greening your supply chain', *Journal of Cleaner Production*, Vol.40, pp. 93-100.

Eltayeb, T. K., Zailani, S. and Ramayah, T. (2011) 'Green supply chain initiatives among certified companies in Malaysia and environmental sustainability: investigating the outcomes', *Resources, conservation and recycling*, Vol.55, No.5, pp.495-506.

Epstein, M. J. (2008) *Making sustainability work: Best practices in managing, and measuring corporate social, environmental, and economic impacts*, Sheffield, UK: Greenleaf Publishing Limited.

Esty, D. C. and Simmons, P. (2011) *The green to gold business playbook: how to implement sustainability practices for bottom-line results in every business function*, NJ, USA: John Wiley and Sons.

—— and Winston, A. S. (2009) *Green to gold: How smart companies use environmental strategy to innovate, create value, and build competitive advantage*, NJ, USA: John Wiley and Sons, Inc.

Eweje, G. (2006) 'Environmental costs and responsibilities resulting from oil exploitation in developing countries: The case of the Niger Delta of Nigeria', *Journal of Business Ethics*, Vol. 69, No.1, pp. 27-56.

Fresner, J. (1998) 'Cleaner production as a means for effective environmental management', *Journal of Cleaner Production*, Vol.6, Issue 3, pp. 171-179.

Frosch, R. A. and Gallopoulos, N. E. (1989) 'Strategies for manufacturing', *Scientific American*, Vol.261, No. 3, pp. 144-152.

Galeazzo, A., Furlan, A. and Vinelli, A. (2014) 'Lean and green in action: interdependencies and performance of pollution prevention projects', *Journal of Cleaner Production*, Vol.85, pp. 191-200.

Gertsakis, J., Ryan, C. and Lewis, H. (1997) *A guide to EcoReDesign: improving the environmental performance of manufactured products*, Melbourne, Australia:Centre for Design, Royal Melbourne Institute of Technology.

Glavič, P. and Lukman, R. (2007) 'Review of sustainability terms and their definitions', *Journal of Cleaner Production*, Vol.15, No.18, pp. 1875-1885.

Grant, D. B., Trautrims, A. and Wong, C. Y. (2013) S*ustainable Logistics and Supply Chain Management: Principles and Practices for Sustainable Operations and Management*, London, UK: Kogan Page Publishers.

Green Jr, K. W., Zelbst, P. J., Meacham, J. and Bhadauria, V. S. (2012) 'Green supply chain management practices: impact on performance', *Supply Chain Management: An International Journal*, Vol.17, No.3, pp. 290-305.

Guide Jr, V. D. R. and Van Wassenhove, L. N. (2009) 'The evolution of closed-loop supply chain research', *Operations Research*, Vol.57, No.1, pp. 10-18.

Gupta, M. C. (1995) 'Environmental management and its impact on the operations function', *International Journal of Operations and Production Management*, Vol.15, No.8, pp. 34–51.

Haden, P. S. S., Oyler, J. D. and Humphreys, J. H. (2009) 'Historical, practical, and theoretical perspectives on green management: an exploratory analysis', *Management Decision*, Vol.47, No. 7, pp. 1041–1055.

Handfield, R. B., Melnyk, S. A., Calantone, R. J. and Curkovic, S. (2001) 'Integrating environmental concerns into the design process: the gap between theory and practice', *Engineering Management, IEEE Transactions on*, Vol.48, No. 2, pp. 189–208.

Hart, S. L. (1997) 'Beyond greening: strategies for a sustainable world', *Harvard Business Review*, Vol. 75, No. 1, pp. 66–77.

Hervani, A. A., Helms, M. M. and Sarkis, J. (2005) 'Performance measurement for green supply chain management', *Benchmarking: An International Journal*, Vol.12, No. 4, pp. 330–353.

Holt, D. and Ghobadian, A. (2009) 'An empirical study of green supply chain management practices amongst UK manufacturers', *Journal of Manufacturing Technology Management*, Vol.20, No.7, pp. 933–956.

Hughes, S. B., Anderson, A. and Golden, S. (2001) 'Corporate environmental disclosures: are they useful in determining environmental performance?', *Journal of Accounting and Public Policy*, Vol. 20, No. 3, pp. 217–240.

Klassen, R. D. and Vachon, S. (2003) 'Collaboration and evaluation in the supply chain: The impact on plant-level environmental investment', *Production and Operations Management*, Vol.12, No.3, pp. 336–352.

Kleindorfer, P. R., Singhal, K. and Wassenhove, L. N. V. (2005) 'Sustainable operations management7', *Production and Operations Management*, Vol.14, No.4, pp. 482–492.

Kolk, A. (2005) 'Environmental reporting by multinationals from the triad: Convergence or divergence?', *Management International Review*, Vol.45, No. 1, pp. 145–166.

Kopnina, H. and Blewitt, J. (2015) *Sustainable business: Key issues in environment and sustainability*, New York, NY: Routledge.

Larson, A. L. (2000) 'Sustainable innovation through an entrepreneurship lens', *Business Strategy and the Environment*, Vol.9, No.5, pp. 304–317.

McDonough, W. and Braungart, M. (2010) *Cradle to cradle: Remaking the way we make things*, New York, NY: North Point Press, MacMillan.

———, ———, Anastas, P. T. and Zimmerman, J. B. (2003) 'Peer reviewed: applying the principles of green engineering to cradle-to-cradle design', *Environmental Science and Technology*, Vol.37, No.23, pp. 434–441.

Melnyk, S. A., Sroufe, R. P. and Calantone, R. (2003) 'Assessing the impact of environmental management systems on corporate and environmental performance', *Journal of Operations Management*,

Vol.21, No.3, pp. 329-351.

Molina-Azorin, J. F., Claver-Corte´s, E., Pereira-Moliner, J. and Tarı´, J. J. (2009) 'Environmental performance and firm performance: an empirical analysis in the Spanish hotel industry', *Journal of Cleaner Production*, Vol.17, pp. 516-524.

Morrow, D. and Rondinelli, D. (2002) 'Adopting corporate environmental management systems: motivations and results of ISO 14001 and EMAS certification', *European Management Journal*, Vol.20, No. 2, pp. 159-171.

Nikolaou, I. E., Evangelinos, K. I. and Allan, S. (2013) 'A reverse logistics social responsibility evaluation framework based on the triple bottom line approach', *Journal of Cleaner Production*, Vol.56, pp. 173-184.

Oberhofer, P. and Dieplinger, M. (2014) 'Sustainability in the transport and logistics sector: lacking environmental measures', *Business Strategy and the Environment*, Vol.23, No.4, pp. 236-253.

Parmar, B. L., Freeman, R. E., Harrison, J. S., Wicks, A. C., Purnell, L. and de Colle, S. (2010) 'Stakeholder theory: The state of the art', *The Academy of Management Annals*, Vol.4, No.1, pp. 403-445.

Peng, Y.-S. and Lin, S.-S. (2008) 'Local responsiveness pressure, subsidiary resources, green management adoption and subsidiary's performance: evidence from Taiwanese manufactures', *Journal of Business Ethics*, Vol.79, No.1-2, pp. 199-212.

Porter, E. M. and van der Linde, C. (1995) 'Green and competitive: ending the stalemate', *Harvard Business Review*, Vol.73, No. 5, pp. 120-134.

Potoski, M. and Prakash, A. (2005) 'Covenants with weak swords: ISO14001 and facilities' environmental performance', *Journal of Policy Analysis and Management*, Vol.24, No.4, pp. 745-769.

Rao, P. (2004) 'Greening production: a South-East Asian experience', *International Journal of Operations and Production Management*, Vol.24, No.3, pp. 289-320.

―― and Holt, D. (2005) 'Do green supply chains lead to competitiveness and economic performance?', *International Journal of Operations and Production Management*, Vol.25, No.9, pp. 898-916.

Rogers, D. S., Tibben-Lembke, R. S. and Council, R. L. E. (1999) *Going backwards: reverse logistics trends and practices*, Vol.2, Pittsburgh, PA: Reverse Logistics Executive Council.

Rusinko, C. A. (2007) 'Green manufacturing: an evaluation of environmentally sustainable manufacturing practices and their impact on competitive outcomes', *Engineering Management, IEEE Transactions on*, Vol.54, No.3, pp. 445-454.

Sarkis, J. (2003) 'A strategic decision framework for green supply chain management', *Journal of Cleaner Production*, Vol.11, No.4, pp. 397-409.

――, Helms, M. M. and Hervani, A. A. (2010) 'Reverse logistics and social sustainability', *Coporate Social responsibility and Enviormental Management*, Vol.17, pp. 337-354.

Saur, K., Donato, G., Flores, E. C., Frankl, P., Jensen, A. A., Kituyi, E., . . . Tukker, A. (2003) 'Draft final report of the LCM definition study',*UNEP/SETAC Life Cycle Initiative*.

Seuring, S. (2004) 'Industrial ecology, life cycles, supply chains; differences and interrelations', *Business Strategy and the Environment*, Vol.13, No.5, pp. 306–319.

Soylu, K. and Dumville, J. (2011) 'Design for environment: The greening of product and supply chain', *Maritime Economics and Logistics*,Vol.13, No.1, pp. 29–43.

Srivastava, S. K. (2007) 'Green supply-chain management: A state-of-the-art litrature review', *International Journal of Management Review*, Vol.9, No.1, pp. 53–80.

Stapleton, P. J., Glover, M. A. and Davis, S. P. (2001) 'Environmental managment systems: an implementation guide for samll and medium-sized organizations', *Environmental managment systems: an implementation guide for samll and medium-sized organizations*, NSF.

Steger, U. (2000) 'Environmental management systems: empirical evidence and further perspectives', *European Management Journal*, Vol.18, No.1, pp. 23–37.

Steurer, R., Langer, M. E., Konrad, A. and Martinuzzi, A. (2005) 'Corporations, stakeholders and sustainable development I: a theoretical exploration of business—society relations', *Journal of Business Ethics*, Vol. 61, pp.263–281.

The United States Environmental Protection Agency (2014) Life-Cycle Assessment, Retrieved 9 April 2015, Available at, http://www.epa.gov/sustainability/analytics/life-cycle.htm

Tsoulfas, G. T. and Pappis, C. P. (2006) 'Environmental principles applicable to supply chains design and operation', *Journal of Cleaner Production*, Vol.14, No.18, pp. 1593–1602.

WBCSD (1996) *Eco-efficiency and cleaner production: charting the course to sustainability*, Geneva, Paris: WBCSD (World Business Council for Sustainable Developrnent) and UNEP (United Nations Environment Programme), Available at
http://oldwww.wbcsd.org/plugins/DocSearch/details.asp?MenuId=ODUandClickMenu=RightMenuanddoOpen=1andtype=DocDetandObjectId=MzAx,
Accessed April 28th, 2015.

Wolf, J. (2014) 'The relationship between sustainable supply chain management, stakeholder pressure and corporate sustainability performance', *Journal of Business Ethics*, Vol.119, No.3, pp. 317–328.

Zsidisin, G. A. and Siferd, S. P. (2001) 'Environmental purchasing: A framework for theory development', *European Journal of Purchasing and Supply Chain Management*, Vol.7, pp. 61–73.

4 Green Management
——Environmental-Financial Performance Nexus and Dimensions of Innovation

Keiko Zaima
Kyoto Sangyo University

[Abstract]

Corporate green management has shifted from performing activities as a good corporate citizen to performing those that also serve a company's self-interest. The objectives of this study are twofold—to give a brief review of the empirical research on the environmental-financial performance nexus and to outline some aspects of a company that were reshaped through green management. In particular, the following four themes are discussed in this paper. Firstly, the empirical research results on the environmental-financial performance nexus are predominantly positive. Secondly, the roles of top managements, created values, and stakeholder relationships can be reshaped by green management. Thirdly, green management can involve three dimensions of innovation. Finally, the spillover effect, as an instance of open innovation, has potential to create a green society.

Keywords : green management, environmental performance, financial performance, roles of top management, creating green value, stakeholder relationship, innovation, innovation diffusion,

1. Introduction

Corporate social responsibility has changed from performing activities as a good corporate citizen to managing business in order to solve a certain social issue related with the company's activity on the basis of its economic self-interest. The meaning of investment in environmental responsibility has also been shifting from costs of avoiding negative environmental impacts to financial returns from bringing positive effects to the company, its shareholders, and its other stakeholders.

Thus, the greening of a business can be said as a set of initiatives that focus on mitigating a company's environmental impacts through its operations and management. The purpose of the green management session of the Japan Forum on Business and Society (JFBS) Conference in 2014 was to discuss the following three questions[1]. — whether experience actually supports the existence of a business case for environmental responsibility, how strong the evidence is, if any, and how the pursuit for a business greening may reshape a company's actions.

As to the first question, excellent business cases were presented at the session with the experiences of Lush Japan Co. Ltd. and Suntory Holdings Ltd, as well as top-scoring Japanese companies in the annual Environmental Management Survey conducted by Nikkei Inc. Thus, this study focuses more on questions 2 and 3. The study's objectives are twofold—with regard to question 2, to show some important insights through a brief review of research on the environmental-financial performance nexus, and with regard to question 3, to outline some insights about how green management can change a company and to suggest three dimensions of innovation in green management: or more specifically, how green management can reshape a company's actions.

The rest of this paper is organized as follows. Section 2 provides an overview of the relationship between the environmental and financial performances through a brief review of empirical research. Section 3 reconsiders the concepts of green management and extracts four insights about how a company reshapes its actions. On the basis of the four insights, Section 4 suggests three dimensions of innovation in green management. Finally, Section 5 presents the study's conclusions.

2. Environmental-financial performance nexus

There is a significant amount of empirical research on the environmental-financial nexus, including some review studies. Molina-Azorín et al. (2009) reviewed 32 studies that had been conducted until 2008 and showed some methodological differences between the studies. Environmental performance variables are divided into two categories those that represent the state of a company's environmental management and process- or product-driven initiatives and those that represent aspects of environmental performance, such as resource consumption and emission of pollutants, toxic chemicals, and so on. Some studies used lagged environmental management variables, because environmental performance cannot be assumed to have an immediate impact on financial performance. The 32 studies used different variables to represent financial performance, namely stock market return, stock price, and profits-related variables such as return on assets, return on sales, and return on equity, as well as different analysis methods, namely regression, event study, structural equation modeling, and correlation analysis. Despite of the above differences in methodology, Molina-Azorín et al. (2009) found a predominantly significant positive relationship between environmental and financial performances, although the results of those studies were mixed from negative to positive nexus. The studies they examined include two-way interactions between environmental and financial performance, with most studies focusing on the effects of environmental performance on financial performance, rather than vice versa.

Horváthová (2010) conducted a meta-analysis of 64 outcomes of 37 empirical studies that were published during the period from December 2008 to February 2009. According to her, 35 of the 64 outcomes showed a positive relationship between the environmental and financial performances, 10 showed a negative relationship, and the remaining 19 did not show significant results. She insisted that the differences in the results came from differences in research methods or country features. Outcomes from correlation coefficient and portfolio studies tended to show a negative relationship, although those studies used multiple regression and panel data analyses. Horváthová (2010) focused on the differences in law systems and divided the research into two categories— that on civil law countries, such as Germany and France, and that on common law countries such as the UK, USA, and Canada. She found that the latter research showed positive results for the environmental-financial performance nexus.

Iwata and Okada (2011) examined the effects of environmental performance on financial performance using data for Japanese manufacturing firms from 2004 to 2008. They found different results on the relationship between environmental issues and performance. They found a significant positive relationship between the environmental and financial performances with regard to greenhouse gas and an insignificant positive relationship with regard to waste emission. They also found different results between industries, which they divided into two categories—clean and dirty—according to precedent studies by other researchers. "Dirty" industries included pulp and paper, chemical, pharmaceutical, rubber product, iron and steel, nonferrous metal, and metal product industries, while "clean" industries included food, textile and apparel, oil and coal product, glass and ceramic product, machinery, electric appliance, transportation equipment, and other industries. Iwata and Okada (2011) clarified that clean industries showed a significant positive environmental-financial performance nexus, while dirty industries showed an insignificant nexus.

González-Benito et al. (2006) investigated the determinant factors of environmental proactivity through a literature review. Environmental proactivity is defined as "the voluntary implementation of practices and initiatives aimed at improving environmental performance"[2]. They found three types of determinant factors—first, stakeholder pressure as a main factor; second, a factor that represents the level of pressure the top management perceives from stakeholders and the attitudes of the top management on environmental management and strategy; and third, is how strong the pressure is. These factors were related with company features, like size, value chain position, degree of internationalization, industry, and location.

Although the above four studies are analyses on large enterprises, there are also several researches on small and medium-sized enterprises (SMEs). For example, Aragón-Correa et al. (2008) conducted an empirical study on 108 SMEs in the automotive repair industry in Southern Spain. They found that environmental strategies in SMEs range from reactive regulatory compliance to proactive pollution prevention and environmental leadership. They also found that the most proactive SMEs in environmental practices exhibited a significantly positive financial performance.

Zaima (2008) conducted an empirical study on Japanese SMEs and found two paths to improving environmental performance. On the one hand, when a company is a subcontractor, or when market circumstances are severe, stakeholder pressure boosts the company's environmental performance. On the other hand, when a company has strong sales, or when market circumstances are not severe, a strong economy boosts the company's environmental performance. In both cases, knowledge support plays an important role.

To summarize the above review, the answer to the second question of how strong the evidence is on the environmental-financial performance nexus is as follows. Although the existing empirical results include both a positive and a negative nexus, and significant and insignificant results, the findings show a predominantly positive relationship between the environmental and financial performances. Most results showed that environmental performance affects financial performance, although some results also show vice versa. The differences in the results come from differences in variables, industries, research methods, and environmental issues. Stakeholder pressure can be regarded as a significant determinant factor of proactive green management, although there are several internal and external factors for proactive green management.

3. Changes through green management

In this section, the question of how green management can reshape a company's actions is discussed. In particular, three viewpoints from the basics of corporate management are described the roles of top management, added values created through corporate management, and a company's relationships with stakeholders.

3-1. Reshaping roles of top management

There is no common definition of green management, in existing literature reviews. Pane Haden et al. (2009) provided a comprehensive definition of green management as "the organization-wide process of applying innovation to achieve sustainability, waste reduction, social responsibility, and a competitive advantage via continuous learning and development and by embracing environmental goals and strategies that are fully integrated with the goals and strategies of the organization"[3]. As shown in this definition, green management is not a specific activity in a narrow field but involves corporate-wide process. And green management is intended to incorporate perspective of innovation and strategy, rather than simply refer to environmental conservation activities.

In a company, making decisions concerning innovation and strategy is one of the most important roles of the top management. In general, the basic role of the top management is to develop the company's philosophy, strategies, and organization. Each company has its own philosophy, which was established by the founder to clarify the company's purpose and type of business. To realize the philosophy, the top management must determine the direction of the company's strategies, and to implement the strategies, it needs to create the organization and systems in order to motivate employees[4].

Fig. 1 shows the basic roles of the top management and how green management can reshape them. Green management starts from the company's green philosophy, in which the top management declares how the company should reduce its negative impacts on the environment and intends to create positive impacts to the society, based on an assessment of the company's relationships with environmental issues. To realize the green philosophy, the top management should redesign the corporate strategies, which can range from functional ones described as environmental performance goals in factories to business ones related to the development of new green technology. The type of strategy taken depends on the green philosophy. Although a reactive strategy, which involves simply compliance to existing environmental laws and regulations, is possible, a proactive strategy, which involves seeking opportunities to reduce costs and gain benefits, can contribute to the company's financial returns, as shown by empirical research discussed in the previous section. A proactive strategy moves the present business and activities toward environmental consciousness. Green management needs a

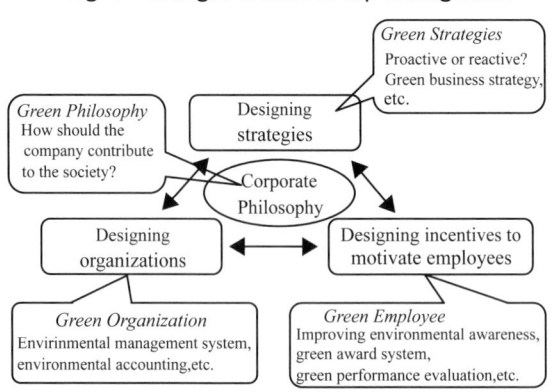

Fig. 1 Changes in roles of top management

```
                                    Green Strategies
                                    Proactive or reactive?
                                    Green business strategy,
                                    etc.
   Green Philosophy    Designing  ↗
   How should the      strategies
   company contribute
   to the society?  ↘   ↓   ↙
                    Corporate
                    Philosophy
              ↙                 ↘
       Designing    ←——→    Designing incentives to
       organizations          motivate employees

    Green Organization          Green Employee
    Envirinmental management    Improving environmental awareness,
    system,                     green award system,
    environmental accounting,   green performance evaluation,etc.
    etc.
```

redesign of the organization by installing some tools, such as an environmental management system, life cycle assessment of products, and environmental audit and accounting. The implementation of green management needs green education to improve the employees' environmental awareness. The top management should redesign incentive systems to motivate the employees toward environmental consciousness.

The most essential aspect of the roles of the top management in green management is to integrate green concepts with original decision making, organization, and management systems in the company.

3-2. Reshaping added values created through corporate management

Corporate management is basically a process in which a company creates added value. The company inputs managerial resources, outputs products through technological transformation, and creates added values to supply products and services that satisfy customer needs[5]. However, negative environmental impacts accompany corporate management processes, using natural and energy resources and emitting wastes and pollutants, as shown in Fig. 2.

Fig. 3 shows an outline of how green management reshapes the corporate management process. In the process of green management, the company installs green technology, implements management methods, supplies green products and services, and performs some green activities in the society. Through such green management activities, negative environmental impacts are reduced and positive impacts are generated. These impacts can be estimated through environmental accounting and can be regarded as "green values." Thus, green management process entails creating added values for the environment within the corporate management process that creates added values for the customer.

Fig. 2 Corporate management process and environmental impacts

```
        Input          Technological      Added Value
                       Transformation
    Human/                                    Output
    Financial/
    Physical    ⇒    Information       ⇒    Product
    Resources         Resources               Service
                Transport                Transport
         ↑              ↓                      ↑
  Natural Resources   Pollitants      Energy    Energy
       Energy   Energy  Wastes,etc     etc.     Wastes
                etc.
```

Negative Impacts on the Environment

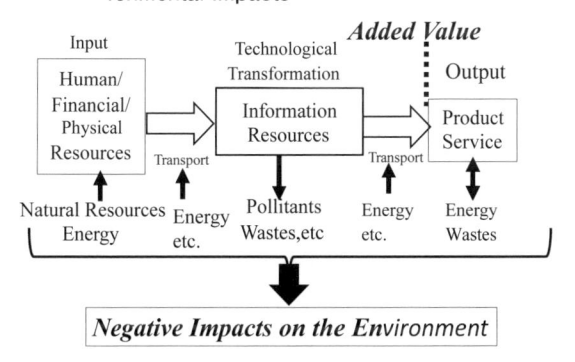

Fig. 3 Changes in added values created through corporate management

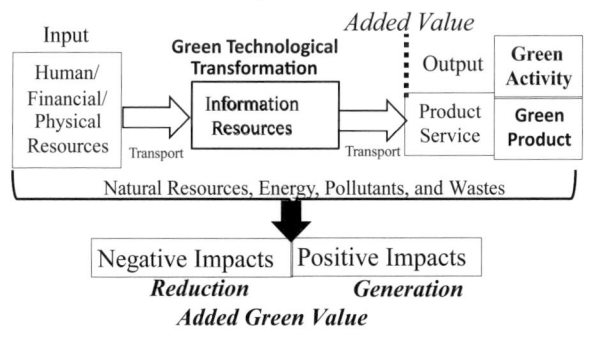

```
        Input          Green Technological    Added Value
                       Transformation
    Human/                                   Output    Green
    Financial/                                         Activity
    Physical    ⇒    Information       ⇒    Product   Green
    Resources         Resources               Service  Product
                Transport                Transport
         Natural Resources, Energy, Pollutants, and Wastes
```

Negative Impacts | Positive Impacts
Reduction | *Generation*
Added Green Value

Recently, the company's purpose has been changing from creating its own added value through corporate management process to creating shared value. The concept of creating shared value is defined by Porter and Kramer (2011) as "policies and operating practices that enhance the competitiveness of a company while simultaneously advancing the economic and social conditions in the communities in which it operates"[6]. For example, Nippon Poly-Glu Co., Ltd. manufactures a specific high-polymer flocculant developed from polyglutamic acid, which is edible, water-soluble, and biodegradable. The chemical promotes the separation of solid particles from a liquid. Nippon Poly-Glu has developed new markets for the Base of Economic Pyramid (BOP) in developing countries[7], in which access to safe water is one of the most serious problems[8]. The company's products not only can purify drinking water, but are also cheap and easy to use. In addition, the company has employed sales-ladies and created new businesses in the developing countries that they entered. The chemical can solve the water safety problem in these countries through a corporate business. In this case, the company's green business or activities can solve a problem in the society, and the company's added green value is consistent with a certain social value. Therefore, the green value can become a shared value in the cases of BOP with green products.

As described above, the most essential aspect of the green management process is to reshape the values that a company can create.

3–3. Reshaping relationships with stakeholders

A company has relationships with various stakeholders in the economy and society. Fig. 4 shows a traditional depiction of a company's relationships with stakeholders, while author suggests Fig. 5 presents a modern depiction by the author of this study of such relationships. In the latter figure, the company conducts its business, facing four types of markets: product and service markets (described as "product/service markets" in the figure), materials markets, financial and capital markets (described as "financial/capital markets" in the figure), and labor markets. The company also faces the central and local governments (described as "government" in the figure), for which regulations and policies must be complied with. The company also needs to be a good corporate citizen in the society and to reply various demands of citizens, NPOs, and NGOs (described as "citizen, NPO/NGO" in the figure). The stakeholders indicated in Fig. 4 can be categorized into the types of markets or entities in society shown in Fig. 5.

Our society and economy depend on the earth[9]. The earth provides us natural resources, which are indispensable for a company's business and activities. Therefore, the earth should also be one of the company's stakeholders shown in Fig. 5[10]. As there exists only one earth, the company should play an important role in the sustainable usage of limited resources, a role that the society has come to expect of the company. In

Fig. 4 Traditional depiction of a company's relationships with stakeholders

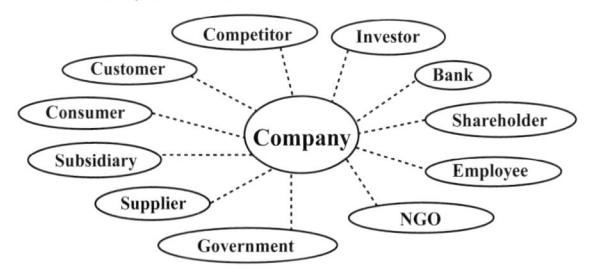

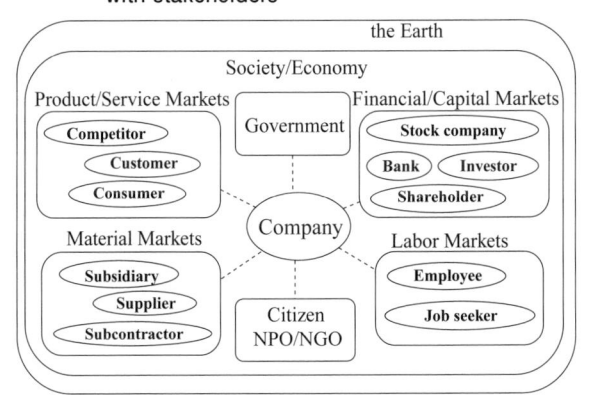

Fig. 5 Modern depiction of a company's relationships with stakeholders

other words, the conditions that must be satisfied by the company have changed.

Fig. 6 outlines the traditional conditions that the company must satisfy in order to exist. The conditions consist of simple action items. The company should provide valuable products to the customers in the market, by differentiating itself and its products from its competitors and their products. It should procure materials necessary for its business and pay to its suppliers. It should raise funds in financial and capital markets. It should employ human capital necessary to its business from the labor market. Finally, it should comply with the laws and regulations established by governments and should respond the demands citizens.

Fig. 7 shows the new conditions imposed on a company to help ensure its existence. The company should provide valuable products not only to the customer but also to the earth and society. It should procure adequate materials for the production of green products. It should disclose both positive and negative information on its financial, environmental, and social performances to investors and other stakeholders. It should ensure that employees' job satisfaction go hand in hand with environmental action. The company can cooperate with governments and NPOs to create a more environmentally conscious society. Through its business and green management activities, the company must create good relationships with its stakeholders. Thus, green management can re-

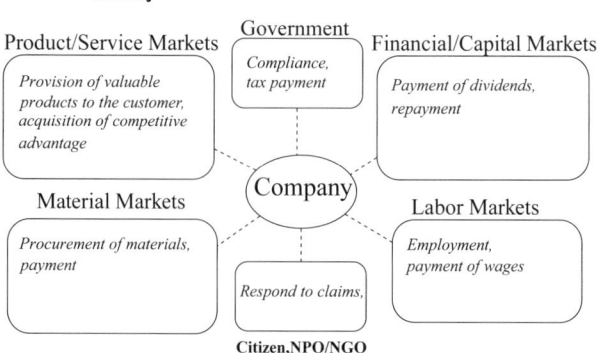

Fig. 6 Traditional conditions that the company should satisfy

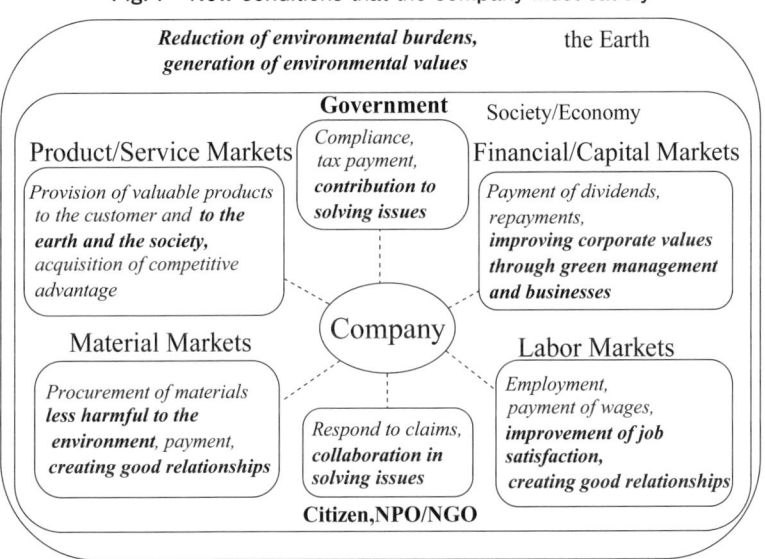

Fig. 7 New conditions that the company must satisfy

Reduction of environmental burdens, generation of environmental values — the Earth

Product/Service Markets
*Provision of valuable products to the customer and **to the earth and the society**, acquisition of competitive advantage*

Government
Compliance, tax payment, contribution to solving issues

Society/Economy

Financial/Capital Markets
Payment of dividends, repayments, improving corporate values through green management and businesses

Material Markets
*Procurement of materials **less harmful to the environment**, payment, creating good relationships*

Company

Respond to claims, collaboration in solving issues

Labor Markets
Employment, payment of wages, improvement of job satisfaction, creating good relationships

Citizen, NPO/NGO

shape relationships between the company and its stakeholders, and those changes can reduce environmental burdens and generate environmental values.

4. Dimensions of innovation concerning green management

As mentioned above, green management can reshape the top management's decision making, and the values that are created through corporate management process and the company's relationship with stakeholders. In a company that has adopted green management, activities may change and new items may be introduced. Such changes can be considered innovations in the company. There are several definitions of innovation. Although the term innovation generally refers to technological innovation, Schumpeter's definition of innovation refers to the development of new products, production methods, markets, and raw materials, the construction of a new organization, and a new combination of these elements. Rogers (2003), who pioneered innovation diffusion studies, defined innovation as "an idea, practice, or project that is perceived as new by an individual or other unit of adoption"[11].

Thus, there are three degrees of newness of innovation. If a certain technology is developed for the first time in the world, it is completely new. This case is the first degree of newness. When a new technology is applied by a company in a certain industry for the first time after the technology was developed in another industry, it is not completely new in general but is new for that industry. This case is the second degree of newness, which applies to a limited range of cases, such as in a country, industry, and community. Technological innovation and Schumpeter's innovation involve these two degrees of newness. Finally, when many companies have already adopted a certain technology, the technology is not considered "new" in general, but it is so for the company that adopts the technology for the first time. Thus, the third degree of newness applies to a person or organization that adopted a certain technology for the first time, even when the technology is already well-known in general. Roger's innovation includes this third degree of newness.

Green management has three dimensions of innovation. The first dimension is Rogers's definition of innovation. Green management concepts and activities are not entirely new these days. However, when a company introduces green management into its organization for the first time, it is new for that company. The second dimension is innovation in new technology, products, production methods, markets, raw materials, organizations, and the combinations of these elements, which are developed for mitigating environmental impacts in the green management process. Such innovation refers to the general definition of innovation or Schumpeter's definition. A company has many options for proactive green management, and its choice depends on the kind of innovation it wants.

The third dimension of green management is the diffusion of green management innovation, which involves the first and second dimensions of green management. In the previous section, three points that green management can reshape were described. The fourth point that can be reshaped is the spillover effect of greening. For example, clause 4.4.6 of operational control in ISO 14001, an international certification of environmental management systems, involves not only establishing, implementing, and maintaining procedures related to the identified significant aspects of the company's products, but also communicating applicable procedures and requirements to suppliers and contractors[12]. To comply with this clause, the certified company asks its suppliers or contractors to complete an environmental audit or questionnaire concerning their environmental activities. Through this process, green management diffused. However ISO 14001 has not diffused green management enough to include, for example, SMEs or developing countries. Therefore, in the next stage of green management, leading companies must be asked how they can contribute to diffusing green management innovation in the supply chain, industry or community. Based on the categories of stakeholders, as shown in Fig. 5, there are six directions that prompt a company to take action as a change agent for innovation diffusion, shown in Fig. 8.

Diffusion of green management innovation can create a sustainable society, where the society's activities and businesses and the economy are greened and both the environmental and financial performances in the company and society are improved. Such

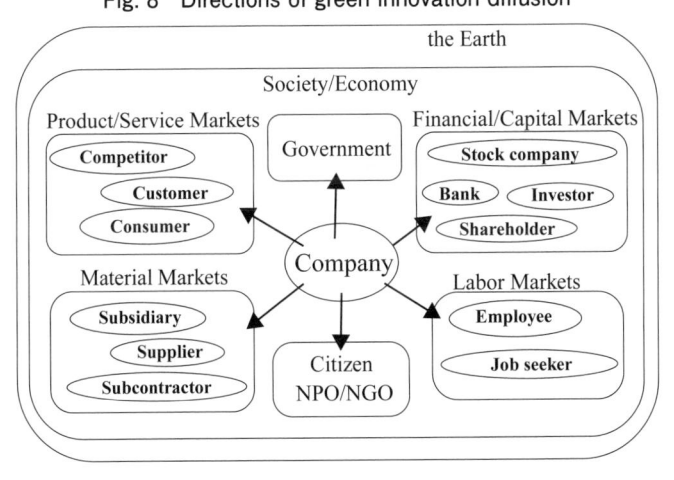

Fig. 8 Directions of green innovation diffusion

a situation can be regarded as social innovation. As it is not easy for a single company to influence others as a leader of green innovation diffusion, a certain kind of multi-stakeholder platform, shown in Fig. 9, should be introduced in the society[13]. Through the platform, stakeholders share green vision and concepts, have dialogue, and assist in knowledge -or technology-related issues. For example, SMEs can acquire useful information on green technology, management methods, and supportive programs provided by governments, that can complement the SME's generally limited managerial and other resources. Such a process can bring open innovation to SMEs. Open innovation is defined by Chesbrough (2003) as "a distributed innovation process based on purposively managed knowledge flows across organizational boundaries, using pecuniary and non-pecuniary mechanisms in line with the organization's business model"[14]. Open innovation refers to the use of not only inside knowledge but also outside knowledge for innovation. Who can take the initiative to implement open innovation for green management diffusion is an important issue for further research and the society.

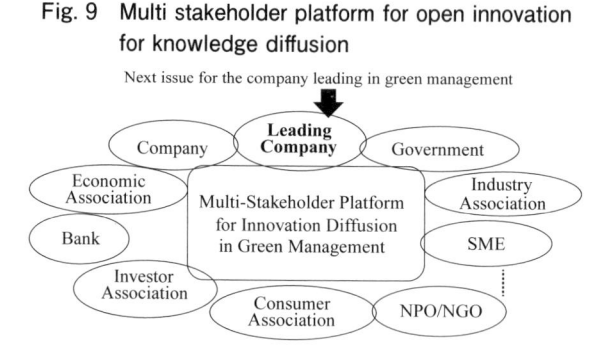

Fig. 9 Multi stakeholder platform for open innovation for knowledge diffusion

5. Concluding remarks

Three questions were posed at the green management session of the JFBS Conference in 2014—first, whether experience actually supports the existence of a business case for environmental responsibility; second, how strong the evidence is, if any; and third, how the pursuit of business greening can reshape a company's actions. This study focused on questions 2 and 3. In response to question 2, an overview of the relationship between the environmental and financial performance was given through a brief review of empirical research. In response to question 3, three dimensions of innovation in green management were suggested, based on a reconsideration of green management concepts and four aspects that a company can reshape through green management. This section presents the study's concluding remarks.

Firstly, the results of existing empirical studies on the environmental-financial performance nexus are predominantly positive, indicating a strong connection between the two types of performances.

Secondly, the roles of top managements, values created through corporate management, and a company's relationships with stakeholders can be reshaped through the green management process. In addition, such changes can introduce innovation to the company.

Thirdly, green management can involve three dimensions of innovation. New ideas, technologies, products, and markets can be developed through green management activities. Reshaping the roles of top management, values created by the company, and the company's relationships with stakeholders can change the corporate management style, even if the green management concepts and technologies are already well-known. Meanwhile, green management can be considered innovation to the company that implements it for the first time—this idea is Rogers's definition of innovation.

Finally, the spillover effect to other companies is another significant aspect of green management. A multi-stakeholder platform can play an important role in diffusing green management innovation. Implementing initiatives for innovation diffusion is a requirement for a company to lead in the next stage of green management.

I acknowledge the financial support of the JSPS KAKENHI Grant-in-Aid for Scientific Research (C) (Grant Number 26340124).

(1) The organizer of the conference presented the questions in advance to the speakers.
(2) See González-Benito et al. (2006) p. 88.
(3) See Pane Haden et al. (2009) p. 1052.
(4) See, for example, Sakashita (2000) pp. 3-14.
(5) See, for example, Itami and Kagono (2003) p.3.
(6) See Porter and Kramer (2011) p.6.
(7) See, for example, Zaima (2011) and the short report titled "Nippon Poly Glu: drinkable water from anywhere" available through the following link:
http://dwl.gov-online.go.jp/video/cao/dl/public_html/gov/pdf/hlj/20140201/10-11.pdf,
(Access date: May 21, 2015)
(8) See UNICEF/WHO (2012).
(9) Senge et al. (2008) showed a containment relationship between the economy, society, and the earth. See Fig. 8.1 in Senge et al. (2008).
(10) Tanimoto (2004) indicated that the earth (the environment) can be treated as a stakeholder of the company. See Fig. 2-2 in Tanimoto (2004) p. 37.
(11) See Rogers (2003) p. 12.
(12) See, for example, Kurosawa (2005) p. 91.
(13) See, for example, the European Commission et al. (ed.) (2012) for the actions of the EU Multi-stakeholder Forum on CSR.
(14) Open innovation is a concept provided by Chesbrough (2003). The concept can be explored in SMEs. Brunswicker and van de Brande (2014) reviewed existing research on open innovation in SMEs.

<References>

Aragón-Correa, J. A., N. Hurtado-Torres, S. Sharma, and V. J. García-Morales (2008) "Environmental strategy and performance in small firms: A resource-based perspective," *Journal of Environmental Management*, Vol.86, pp. 88-103.

Brunswicker, S. and V. van de Brande (2014) "Exploring Open Innovation in Small and Medium-sized Enterprises," pp. 135-156, in Chesbrough, H. W., W. Vanhaverbeke and J. West (eds.) (2014) *New Frontiers in Open Innovation, Oxford University Press*.

Chesbrough, H. W. (2003) *Open Innovation: The New Imperative for Creating and Profiting from Technology*, Harvard Business School Corporation.

European Commission et al. (ed.) (2012) *EU Multi-Stakeholder Forum on Corporate Social Responsibility: CSR EMS Forum*, Dictus Publishing.

González-Benito, J. and Ó. González-Benito (2006) "A review of determinant factors of environmental proactivity," *Business Strategy and the Environment*, Vol.15, pp. 87–102.

Horváthová, E. (2010) "Does environmental performance affect financial performance? A meta-analysis," *Ecological Economics*, Vol.70, pp. 52–59.

Itami, H. and T. Kagono (2003) *Zeminaru keieigaku (Seminar about business administration) (3rd ed.)*, Nihon Keizai Shinbunsha (written in Japanese).

Iwata, H. and K. Okada (2011) "How does environmental performance affect financial performance? Evidence from Japanese manufacturing firms," *Ecological Economics*, Vol. 70, pp. 1691–1700.

Kurosawa, S. (2005) *ISO 14001 Yasashii Guide Book (An Understandable Guide Book for ISO 14001)*, Nakanishiya Shuppan (written in Japanese).

Molina-Azorín, J. F., E. Claver-Cortés, M. D. López-Gamero, and J. J. Tarí (2009) "Green management and financial performance: A literature review," *Management Decision*, Vol. 47, pp. 1080–1100.

Pane Haden, S. S., J. D. Oyler and J. H. Humphreys (2009) "Historical, practical, and theoretical perspectives on green management: An exploratory analysis," *Management Decision*, Vol. 47, pp. 1041–1055.

Porter, M. E. and M. R. Kramer (2011) "Creating shared value: How to reinvent capitalism and unleash a wave of innovation and growth," *Harvard Business Review*, Vol. 2011 January-February, pp. 2–17.

Rogers, E. M. (2003) *Diffusion of Innovations (5th ed.)*, New York: Free Press.

Sakashita, A. (2000) *Keieigaku eno Shoutai (An introduction to business administration) (revised ed.)*, Hakutou shobou (written in Japanese).

Senge, P., B. Smith, N. Kruschwitz, J. Laur and S. Schley (2008) *The Necessary Revolution: How Individuals and Organizations Are Working Together to Create a Sustainable World*, Nicholas Brealey Publishing.

Tanimoto, K. (ed.) (2004) *CSR Keiei (CSR and Stakeholder)*, Chuoukeizaisha (written in Japanese).

UNICEF/WHO (2012) *Progress on Drinking Water and Sanitation.*

Zaima, K. (2008) "Chushou-kigyou no kankyou keiei suisin no jouken ni kansuru jisshou bunseki: Kikai kinzoku-gyou to purasuchikku-kakou-gyou no keisu (An empirical study on the conditions to promote environmental management of small and medium sized enterprises)," *Shakai-Keizai Sisutemu (Social and Economic Systems)*, Vol.29, pp. 67–76 (written in Japanese with English abstract).

5 CSR and Corporate Reputation
——Towards Effective Strategy Approaches![1]

Tobias Bielenstein
Managing Partner Branding-Institute CMR AG

[Abstract]

CEOs worldwide state that their prime motivator to conduct CSR measures is to strengthen their organizations' reputation–but only a few seem to successful deliver on this goal. This paper addresses the question of the requirements to strategic approaches in order to strengthen corporate reputation by CSR which in consequence asks for the business case of CSR. It argues on a theoretical perspective that successful strategies have to take into account the full range of reputation dimensions, which implies also considering all relevant stakeholders. This argumentation is transferred to a matrix serving as an analytic tool and in a later stage as a guiding instrument in strategy development. The matrix is applied as an analytical tool in a brief case study about the German outdoor company Vaude.

Keywords : CSR, Sustainability, Corporate Reputation, CSR Management, Reputation Management, Strategy, Stakeholder, Stakeholder Theory, Stakeholder Management, Case Study

1. Introduction

69% of CEOs worldwide state that strengthening brand, trust and reputation is their prime motivator to take action on sustainability issues (UN Global Compact and Accenture, 2013, p. 37) This goal ranks even higher than the goal to employ sustainability measures to raise revenue and reduce costs. But the UN Global Compact study also shows that a majority of the more than 1000 CEOs interviewed struggle to locate and quantify the business value of sustainability and to deliver information on the business case. The study suggests the existence of a gap between a majority of companies with CSR approaches based on philanthropy, compliance, mitigation, and the license to operate and so called leading companies, defined as performing clearly above their industry peers in both sustainability and business performance metrics. These companies utilize sustainability strategies to focus on innovation, growth and new sources of value (ibid., p. 2).

The study also proposes that most companies still struggle to develop and implement strategies that can contribute to progress in the global challenges, i.e. to address the needs of a growing world population under the condition of environmental and resource constraints (ibid., p. 11). They stay within programs of limited and/or incremental impact on sustainability metrics. In sum the study suggest that most companies fail both to connect sustainability activities to their core business and to global challenges.

The findings of the UN Global Compact study seem to be in line with other assessments and studies. E.g. Porter and Kramer state that "the prevailing approaches to CSR are so fragmented and so disconnected from business and strategy as to obscure many of the greatest opportunities for companies to benefit society (Porter and Kramer, 2006, p. 80). Van Maverick (2003, p.103) refers to a study conducted by Ernst & Young. According to it 94 % of 114 companies of the "Global 1000" report that sustainability strategies might result in better financial performance while only 11 % of these companies report that they actually implement them accordingly. A survey conducted by MIT Sloan Management Review in collaboration with Boston Consulting Group shows different numbers only on first sight. According to it 48 % of the respondents of the survey report that they changed their business model because of sustainability considerations (Kiron, Kruschwitz et al., 2013, p. 15). But the report leaves open of what kind these changes were and whether they were successful. In addition it isn't disclosed whether the survey is representative or potentially biased because companies with successful sustainability strategies could be more likely to respond in such an online-survey.

In spite of these findings there seems to be a need to introduce effective sustainability strategies. 63 % of the CEOs interviewed in the Global Compact Study are convinced that sustainability will transform their industry within the next five years (UN Global Compact and Accenture, 2013, p. 11). This statement is in consequence a call for action to CEOs to prepare their respective organizations for the transformation anticipated.

This paper aims to address the question of the relationship of Corporate Reputation and CSR with regard to effective strategic approaches. Therefore this paper deals in the first two sections with definitions of CSR and Corporate Reputation respectively. The third section discusses the interrelationship of CSR and Corporate Reputation and takes a look into respective literature. As a conclusion it will be argued in this section from the theoretical perspective that CSR measures have to take into account the whole range of reputation dimensions. This argumentation is transferred to a matrix serving as an analytic tool and in a later stage as a guiding instrument in strategy development. The matrix is applied as an analytical tool in a brief case study about the German outdoor company Vaude.

2. Sustainability and CSR

The number of terms used in the context of business and sustainability is endless: Sustainability, Corporate Sustainability, Corporate Responsibility, Corporate Citizenship, Corporate Social Responsibility, etc. Sometimes they are used synonymous, sometimes with a clear intention to distinguish either whole concepts or different nuances of basically similar concepts. CSR approaches can also be clustered according to political, integrative, ethical and instrumental approaches as well as to regional and national approaches (Dahlsrud, 2008). From the business perspective CSR can also be defined by distinguishing areas of responsibility, i.e. an inner (within the corporation), a middle (within the supply chain) and an outer scope (within society) of responsibility (Hiß, 2006).

As the discussion about definitions of sustainability and CSR cannot be traced here, the widely used definition coined by the Commission of the European Union will serve in this paper as a starting point. In a 2001 communication document the Commission defined CSR as "a concept whereby companies integrate social and environmental concerns in their business operations and in their interaction with their stakeholders on a voluntary basis" (European Commission, 2001, p. 6). In 2011 the European Commission suggested a new definition of CSR as "the responsibility of enterprises for their impacts on society" (European Commission, 2011, p. 6). While both definitions underline the stakeholder orientation of CSR the 2011 definition underlines the expectation that business should have "processes to integrate social, environmental, ethical, human rights and consumer concerns into their business operations and core strategy in close collaboration with their stakeholders" (ibid.).

While Sustainability and CSR are used as synonym in some cases (Dahlsrud, 2008) the "new" EU definition contains only an implicit reference to sustainability when it defines CSR as helping to build trust as a basis of a sustainable business model. In contrast the ISO 26000 definition consistently uses the term sustainable development as a goal of CSR (ISO 2010). From business side it could be argued that the management task of CSR and Sustainability are identical (Loew and Rohde, 2013). But a closer look to the EU Commissions definition shows that CSR actually encompasses more than "just" sustainability. According to the EU Commission CSR covers "at least [covers] human rights, labour and employment practices (such as training, diversity, gender equality and employee health and well-being), environmental issues (such as biodiversity, climate change, resource efficiency, life-cycle assessment and pollution prevention), and combating bribery and corruption. Community involvement and development, the integration of disabled persons, and consumer interests, including privacy, are also part of the CSR agenda" (European Commission, 2011, p. 7). So in addition to the environmental and social dimension usually covered by the term sustainability the EU definition of CSR ads the dimension of corporate governance. This in in accordance with a "broad" understanding of CSR in which Corporate Governance is being defined as an integral part (Schwalbach and Schwerk, 2014, p. 206).

Other than sustainability the term CSR implicates a direct reference to stakeholders. The term "social" points to society, i.e. a broad set of potential stakeholders and the term "responsibility" is a reference to potential addressees of measures to be taken. However the EU Commission's definition of CSR does not distinguish between stakeholders except for the fact that shareholders are mentioned separately (European Commission, 2011, p. 6). Other definitions suggest to differentiate stakeholders in the context of CSR. One approach to do so is provided by Marcel van Marrewijk (2003). While he recommends that a "one solution fits all—definition for CS(R) should be abandoned" (ibid., p. 95) he introduces an option to describe more specifically the development, awareness and motivation level of an organization towards CSR. Based on the "Spiral Dynamics memes" he introduces five "ambition levels" to distinguish particular goals and the scopes of action connected to CR/CSR. According to Marrewijk it is possible to distinguish compliance-driven CS/CSR, profit-driven CS/CSR, caring CS/CSR, synergistic CS/CSR and holistic CS/CSR. While the five stages are pretty self-explaining by their titles they contain the notion of certain stakeholder groups addressed by each of the "ambition levels". E.g. compliance-driven CSR would contribute to "providing welfare to society, within the limits of regulations from the rightful authorities, organizations might respond to charity and stewardship considerations" (ibid., p. 102). This kind of CSR would primarily address governmental and societal stakeholders. Profit-driven CSR is defined as "integration of social, ethical and ecological aspects into business operations and decision-making, provided it contributes to the financial bottom line" (ibid., p. 102). This would mainly address up- and downstream business partners, the financial and investor community accordingly and potentially contradict definitions of CSR that integrate corporate governance, as it is the case in the EU Commission's definition. It is also important to consider that the Spiral Dynamics approach contains the idea, that each "stage" encompasses the foregoing level (Beck and Cowan, 2007). This implicates growing complexity as well as growing input of resources and a growing number of stakeholders addressed at each level. Consequently holistic CSR is defined as follows: "CS(R) is fully integrated and embedded in every aspect of the organization, aimed at contributing to the quality and continuation of life of every being and entity, now and in the future..." (van Marrewijk, 2003, p. 103).

While the concept of Spiral Dynamics itself as well as its application to CSR by van Marrewijk would be worth a closer, critical look, two implications are of interest here. 1. It is possible to distinguish "limited" or focused scopes of CSR activities that could be associated with certain motivations of a corporation, with reference to specific stakeholder groups and respective institutional frameworks. 2. More holistic approaches are accompanied by more complexity e.g. in terms of stakeholders groups addressed and institutional framework needed. By this van Marrewijk's definition helps to understand the high complexity and demanding requirements concerning institutional and governmental efforts for a corporation to establish CSR management in a holistic sense.

3. Corporate Reputation

As is true for CSR also Corporate Reputation is a multidimensional and vague term with many definitions. It is especially difficult to trace back the discussion and to built on existing literature as there are too many different "blueprints" not fitting to each other (Barnett and Pollock, 2012). Probably one the most cited definition for Corporate Reputation is coined by Fombrun and Riel (1997, p. 10): "corporate reputation is a collective representation of a firm's past actions and results that describe the firm's ability to deliver valued outcomes to multiple stakeholders". Barnett, Jermier et al. (2006) put more emphasis on the point that corporate reputation is basically a judgment made by stakeholders and therefore suggest a different definition: "Observers' collective judgments of a corporation based on assessments of the financial, social, and environmental impacts attributed to the corporation over time"(ibid., p. 34). While this definition underlines the notion that reputation is "in the heads of the stakeholders" it contains also a remarkable extension. By defining "financial, social, and environmental impacts", Barnett et al. introduce the idea of certain reputation dimensions that could be understood as a direct reference to sustainability as "financial, social, and environmental" represent the so called "triple bottom line". While this approach was widely accepted it narrowed the definition of reputation and at the same time opened the field for disagreement with this definition (Fombrun, 2012).

Other definitions of corporate reputation either with academic background e.g. Walker (2010), Schwaiger (2004) or with a more popular approach, e.g. Fortune Magazine's "World Most Admired Companies" or "hybrids" combining an academic with popular, PR-oriented approaches, e.g. Reputation Institute's "RepTrak" studies (RepTrack, 2015), all apply certain reputation dimensions. A comparative analysis of eight approaches derives a set of nine reputation dimensions that are most common in current practice (Renner, 2011).

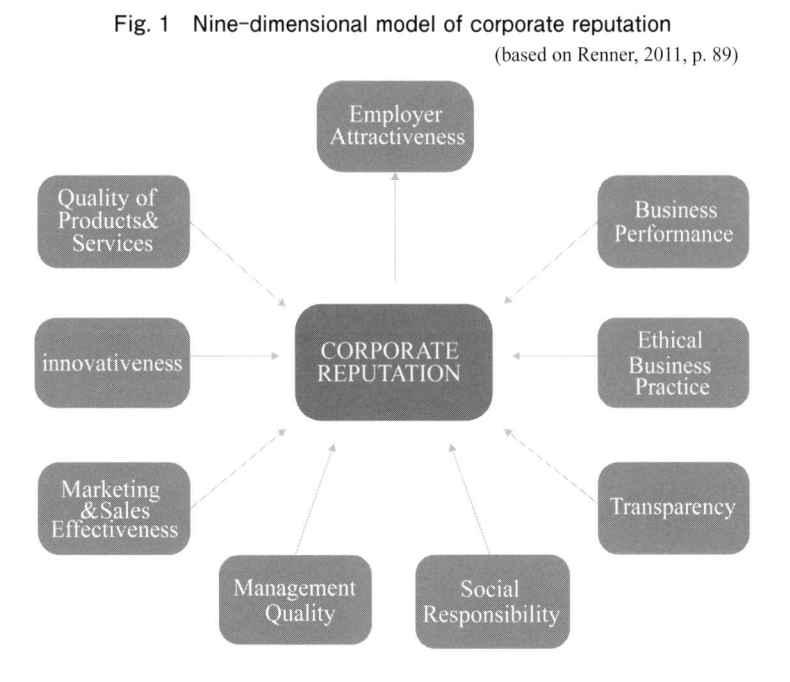

Fig. 1 Nine-dimensional model of corporate reputation
(based on Renner, 2011, p. 89)

These reputation dimensions could be understood both as drivers of reputation in management practice as well as analytical dimensions in order to analyze a corporation's reputation among stakeholders. The reputation dimensions again consist of several indicators that vary for different industries and enterprises. In addition it should be noted that reputation dimensions are not stable. New dimensions emerge and will be established through public discourses and will than serve as criteria against which corporations will be measured by their stakeholders (Kennedy, Chok et al., 2012).

Furthermore the relevance of a reputation dimension should be understood a being stakeholder specific (Renner, 2011). I.e. some dimensions can be of less or no relevance to a specific stakeholder group to evaluate a corporation's reputation while a different stakeholder group would consider the same reputation dimension as most relevant. The study of Renner also shows that certain reputation dimensions associated to CSR may have no reputational effect on certain stakeholder groups (ibid., p. 173).

In corporate practice the question of reputation has gained importance in the past years. It is more and more distinguished from "image" which is seen as a more short-term construct that is not necessarily associated with actual actions of a corporation. In addition image is usually associated with a single persons view while reputation encompasses a collective perception (Einwiller, 2014). Corporate reputation instead is identified as an important precondition for trustful interactions of an organization with its stakeholders. But it is important to note that having a "good reputation" is not the ultimate goal of any corporation. In fact good reputation can aid in environments of limited information, it is a precondition for trust, which is again a precondition for many stakeholders to engage with a corporation, i.e. recommend or buy a product. Furthermore it should be noted that reputation—in differentiation to image—is not derived by communications but on a defined identity of an organization as well as consistent behavior based on this identity, i.e. developing and selling certain products, refraining from selling "bad" or "harmful" products, etc. (Fleischer, 2015).

In sum the relevant points here are that 1. Corporate Reputation is similar to CSR a concept connected to a corporation's set of stakeholders and 2. Corporate reputation consists of several reputation dimensions that can be very diverse with regard to stakeholders, the industry a corporation is belonging to, and the relevance stakeholders ascribe them in interaction with a corporation.

4. Connecting CSR and Corporate Reputation

The notion that CSR contributes to Corporate Reputation is well established in corporations and serves as a standard rationale to take action in CSR. Examples of respective studies were introduced above. But it seems to remain unclear how exactly CSR translates into enhanced reputation of a company and what strategies lead to respective outcomes. As a first step a view into respective literature asks for existing approaches and "mosaic-pieces" to the CSR-Reputation relation.

Most obvious is the fact that CSR could lead to both a positive reputation as well as to the opposite effect i.e. CSR measures having negative impact on corporate reputation. The most familiar term in this context is "Green Washing" which coins cases in which CSR measures are not carried out in substance but remain an "invention of PR" (Frankental, 2001). Connected to "Green Washing" are cases in which CSR measures and the respective reporting attracts attention of critical stakeholders. Possible results of raised awareness are unsubstantial allegations of Green Washing or the risk that stakeholders criticize a corporation "not doing enough" in CSR.

The idea of CSR serving as "reputation insurance" is also well examined. Findings consider that e.g. in product recalls companies with better CSR ratings do better than those without (Minor and Morgan, 2011). This underlines the notion of a positive "reputation account" that could be stocked up in good times. But having a good reputation in a certain area can also provoke strong stakeholder sanction if wrongdoing happens in the same realm. This is shown by Janney and Gove (2011) for the area of corporate governance in a study about the Stock Option Backdating Scandal. Another study has shown that good reputation in CSR could even influence the perception of consumers. According to the study consumers are ready to sort out negative information on a company more easily if it is known to have a good CSR track record (Bhattacharya and Sen, 2004).

However all examples mentioned are based on implicit reputation concepts and therefore remain mosaic-stones with regard to the CSR—Corporate Reputation relation. They also fall short in handling the question regarding strategy development to realize a "reputational return of CSR".

A search for strategy approaches to implement CSR in corporations reveals several examples. A brief view into three prevalent examples is made in order to examine their contribution to the CSR-Corporate Reputation relation:

- **Theory of firm-model**

The theory of firm-model was introduced by McWilliams and Siegel (2001). It suggests that managers can determine an optimal level of resources devoted to CSR activities by conducting a cost/benefit analysis. In this perspective investments in CSR can also contribute to a better differentiation of a company against its competitors. But in this approach reputation seems to be more a means but not an end of CSR activities. An example given by McWilliams and Siegel suggests to invest in CSR advertisements in order to "build or sustain a reputation for quality, reliability, or honesty" (ibid., p. 121). While this seems instrumental on first sight McWilliams and Siegel explain that such a move could also imply investments in R&D, workforce etc. With regard to the question of the CSR-Corporate Reputation relation the theory of firm model bears two shortcomings. First it works only with an implicit concept of Corporate Reputation. Secondly it applies a limited set of stakeholders ("Consumers" and "Others"). Therefore it is too limited with regards to the definition of corporate reputation given above.

- **Shared-Value-Approach**

The Shared Value approach of Porter and Kramer (2011s) is one of the most prevalent approaches in the field of CSR strategies. It received high attention especially among corporate managers but also substantial critics regarding its theoretical and practical foundation (Crane, Palazzo et al., 2014). Regarding the question of CSR and Corporate Reputation the approach contains no substantial contents. It applies an implicit definition of corporate reputation only. In fact Porter and Kramer's definition of reputation seems to be nearer to "corporate image" as defined above. Therefore their critique of CSR programs focusing mostly on reputation and being disconnected from the business case (ibid., p. 16) misses its mark. It provides no further insights to the question of how to strategically implement CSR in order to enhance Corporate Reputation.

・ **Stakeholder Theory**

Edward Freeman's Stakeholder Theory (Freeman, Harrison et al., 2010) is a strategic approach to the management of entire corporations. It basically asserts that management must take into account all relevant stakeholders' views. While this is in line with the stakeholder-based definitions of CSR and Corporate Reputation introduced above Stakeholder Theory raises the claim that it can supersede other concepts. E.g. it states with regard to CSR that "the idea of 'Corporate Social Responsibility' is probably superfluous" (ibid., p. 60) in the sense that CSR should not be separated from business. While the Corporate-Reputation approach introduced above is in part based on Stakeholder Theory, Stakeholder Theory itself lacks an examination of Corporate Reputation. Therefore it also lacks specific information on strategy implementation regarding CSR and Corporate Reputation.

The brief examination of CSR-related literature reveals that despite the notion that CSR contributes to Corporate Reputation there is neither a comprehensive analytical tool nor a ready-to-apply strategic approach available. The "mosaic stones" mentioned above do not add up to a "complete picture" outlining a strategic approach.

5. Towards a strategic approach connecting CSR and Corporate Reputation

Based on these preliminary considerations a draft concept has been developed that serves as the basis to deliver a comprehensive strategic approach connecting CSR and Corporate Reputation. It was developed in practice and is based on the definitions of CSR and Corporate Reputation given above. To apply it in case studies it is drilled down to an analytical tool here. By testing it in case studies it will be further refined and then applied as a framework for strategy development in order to optimize the "reputational return of CSR". The version presented here is simplified. It omits elements that would contribute to further complexity, e.g. indicator sets behind the reputation dimensions.

The first basic consideration is that CSR has to reflect all nine reputation dimensions introduced above. While it could be argued that only three dimensions—social responsibility, transparency and ethical business practices—are directly connected to CSR (see Fig. 2) it is the sum of all dimensions that define a corporation's reputation. Furthermore it is argued here that CSR has to cover by definition all relevant strategic and operational areas of a company. This is e.g. enunciated in the EU definition of CSR enlisting "innovative products, services and business models that contribute to societal wellbeing and lead to higher quality and more productive jobs" (European Commission,

Fig. 2 Nine-dimension model of reputation dimension, according to Renner (2011). Green circles mark reputation dimensions directly associated to CSR

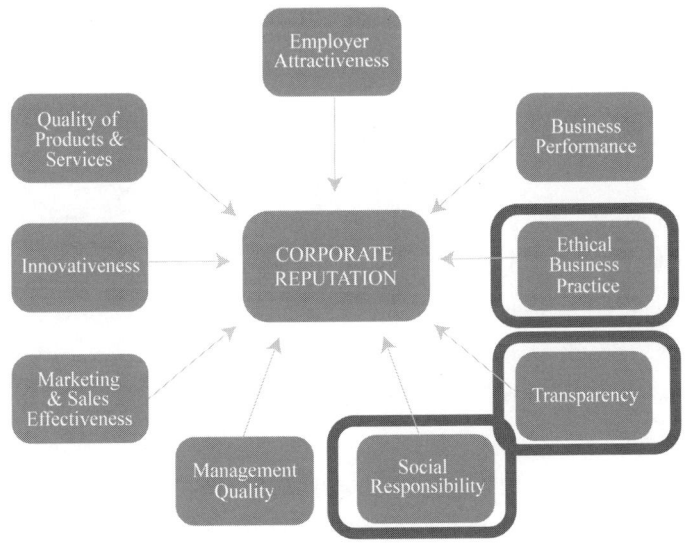

2011, p. 6). As a consequence all reputation dimensions have to be considered in the model.

The second basic consideration is to put all nine reputation dimensions in context to relevant stakeholders of a corporation. This follows the definition of Corporate Reputation as being stakeholder specific. Both considerations are represented in a simple matrix combining reputation dimensions and relevant stakeholder sets. The columns of the table show the nine reputation dimensions. The rows of the matrix consist of stakeholders relevant to the respective corporation, or an entire industry depending on the level of analysis. The fields of the matrix will be filled in with CSR related measures. In addition they show in each field the anticipated reputational outcome regarding a specific stakeholder-group, i.e. endorsement, recommendation, application or purchase. In some cases it would be also appropriate to indicate respective negative outcomes in order to identify reputational risks. It is important to note that these potential outcomes could be different for the same actions: e.g. better environmental performance indicators of a product could result in recommendations by an environmental NGO while it would also result in purchases by consumers.

In an analytical process it is likely the several fields of the matrix would be left void. Many fields in one column would point to a reputation dimension of high relevance. Many completed fields in one row would suggest to a stakeholder group of high influence on reputation. Empty rows and/or columns could indicate strategic and/or operational gaps. If one CSR measure would pop up in different rows/columns this would point to an activity of high relevance for reputation.

While this matrix could remind of one applied in materiality assessments of sustainability it is important to underline the difference between both here. Materiality assessments aim to identify areas in which a company and its stakeholders share respectively disagree on their assessment on sustainability issues. Therefore it is limited to the area of sustainability and to respective assessments and perceptions. The analysis according to the matrix shown in Table 1 asks for activities of an entire company with regard to CSR. This analysis is structured and assigned to nine reputation dimensions. Each ac-

Table 1 Simplified Model of an analytical tool to examine CSR–Corporate Reputation relations with regard to the business case.

	Employer Attractive-ness	Business Perfomance	Ethical Business Practice	Trans-parency	Social Responsi-bilty	Manage-ment Quality	Marketing & Sales Effective-ness	Innova-tiveness	Quality of products & Services
Stake-holder-group A									
Stake-holder-group B									
Stake-holder-group C									
…									

tivity again is assessed with regard to each relevant stakeholder group and respective potential reputational consequences. The latter step is most relevant as it reflects the business case of CSR.

The matrix could be applied both in analytical processes as well as in strategy development. While it looks relatively simple at first glance its complexity becomes immediately clear if it is applied to a major national or international corporation. Already gathering the necessary information could mean the involvement of a high number of corporate departments. Furthermore it could imply new data collections, as necessary information could not be available in all cases etc. When it is applied in developing a strategy the matrix illustrate that this process has to cover the entire company. Therefore the analytical as well as the strategy development process could be less complex in privately held middle scale companies.

The complexity of such an endeavor could also be a hint why many companies fail to realize their reputational return on CSR. Bringing an entire organization on such a path needs continued attention of senior management and long-term commitment. It also points to a potential catch 22 CEOs may end up when introducing CSR measures that should contribute to corporate reputation. While CSR activities of limited scope can be introduced and measured more easily they will not necessarily pay off in higher reputation as they may only address limited numbers of stakeholders and/or reputation dimensions. If then the return on CSR fails to appear additional or more complex CSR measures are unlikely to be introduced. Furthermore it is not easy to track back positive or negative outcomes of such an endeavor directly to the CSR-Reputation strategy and to sort out "external" effects.

6. Brief Case Study[2]──── German Outdoor Company Vaude and the Partnership for Sustainable Textiles

German Outdoor Company Vaude is a producer of outdoor wear and equipment. It is privately held (family owned). CEO is Antje von Dewitz who succeeded her father in 2009 who founded the company. The company has some 500 employees in Germany and some 1000 in Vietnam. As it is part of the textile industry the company relies on global supply chains, which involves typical CSR-issues like labor-and safety standards, logistics etc. In addition the outdoor industry uses chemicals in textiles in order to have their products water- and/or rainproof. This gained critiques of several global environmental NGOs as these chemicals are said to harm the health of both workers producing the apparel as well as the consumers wearing them (Greenpeace, 2015).

In 2008 Vaude set the goal to become Europe's most sustainable outdoor company by 2015. Since then Vaude is both on the radar of many critical NGOs as well as part of media coverage. According to several NGOs, certifiers and media reports the company has become a leader in sustainability and is probably able to reach its goal this year [3]. Vaude established a comprehensive reporting system and was the first company of the outdoor industry to be EMAS eco-certified.

Attached below is an extract of the CSR-Corporate Reputation matrix of Vaude. It shows examples of typical CSR-issues of high relevance for corporate reputation. These examples show up in several fields as well as in identical rows. The goal here is not to present the whole analytical matrix but to illustrate exemplary one strategic decision in terms of CSR and Corporate Reputation.

After the accident in the textile plant in Sabhar 2013 in Bangladesh which let to the death of more than 1.100 workers the German government started an initiative to establish a "Partnership for Sustainable Textiles" bringing together producers, retailers, associations as well as governmental and non-governmental organizations (Partnership for Sustainable Textiles, 2015). Most companies of the German textile industry as well as retailers decided not to become part of the partnership. Despite this Vaude announced in an early stage that it would become member of the partnership (Bauchmüller, 2014).

While it is very likely that Vaude would have chosen to become part of the partnership for reasons of credibility and consistency the CSR-Reputation matrix suggests, that Vaude had no choice but to become member of the partnership: The issue of labor- and safety-standards was of highest relevance for Vaude's reputation. The table shows that these standards are part of two reputation dimensions and relevant to multiple stakeholder groups. With the appearance of a new and powerful stakeholder (German Government) in the same issue area, introducing a relevant multi-stakeholder-platform, it was of at the core of Vaude's corporate reputation to become part of this partnership. In addition Vaude had broad media coverage for this move underlining its claim to be Europe's most sustainable outdoor company. Furthermore it is noticeable that many other companies have become member of the partnership in the meantime proving the importance of this platform in terms of CSR and corporate reputation.

Table 2 Extract of the CSR/Reputation strategy analysis matrix for Vaude with potential stakeholder-specific reputational consequences (bold), simplified representation

	Employer Attractiveness	Business Perfomance	Ethical Business Practice	Transparency	Social Responsibilty	Management Quality	Marketing & Sales Effectiveness	Innovativeness	Quality of products & Services
Customers (Retails etc.)			compliance to labor-/safety-standards | **purchase**	Reporting | **neutral**				No chemicals / more likely to puchase	No chemicals, compliance to labor-/safety-standards | **purchase**
Consumers			compliance to labor-/safety-standards | **purchase**					No chemicals | **purchase**	No chemicals, compliance to labor-/safety-standards| **purchase**
Employers	High attractiveness for family | **motivation/applications of high potentials**		compliance to labor-/safety-standards | **motivation/applications of high potentials**	Reporting					No chemicals, compliance to labor-/safety-standards | **motivation**
NGOs			compliance to labor-/safety-standards | **refrain from blaming/recommendation**	Reporting | **refrain from blaming**				No chemicals | **recommendation**	No chemicals, compliance to labor-/safety-standards | **recommendation**

Business Partners (Suppliers)								
Regulators				Reporting				
NEW: Government			compliance to labor-/safety-standards | **regulation, sanctions**					compliance to labor-/safety-standards | **regulation, sanctions**

7. Outlook

The analysis of literature about CSR and Corporate Reputation suggests a gap in understanding the relation of CSR and Corporate Reputation with respect to strategy development in corporations. The matrix introduced narrows this gap by offering a tool to analyze the CSR-Corporate Reputation relation and to develop respective strategies—and to bring in the business case. Further refinement based on additional experimental case studies is needed to prove whether the approach is sustainable for corporate practice. In addition the tool provides a hint on why many companies fail to gain the reputational return on CSR because of the complexity and the mandatory involvement and commitment of the entire company.

(1) This Paper is based on an introduction given to the breakout session on "Sustainability and Brand & Reputation Management" at the 2014 Annual JFBS Conference. A special thank goes to Prof. Kanji Tanimoto for valuable feedback, to the panelists Takeshi Ohta (Kirin), Hidenori Imazu (Toppan) and David Hessekiel (Cause Marketing Forum) and the participants of the session for comments and feedback at the break-out session as well as to my colleague Dr. Markus Renner for many valuable discussions about the topic.

(2) In addition to the sources cited the base of this case study was derived in a workshop on CSR and transparency which was conducted with senior managers, CMOs and CEOs of several European

outdoor companies at the European Outdoor Summit in Rottach-Egern/Germany in October 2014. It was moderated and co-organized by the author:
http://www.europeanoutdoorsummit.com/index.php/past-events/tegernsee-2014/
(3) E.G.: Clean Clothes Austria (2012) "Firmen Check 2012: Outdoor.", Rank a Brand (2015) "2015 Ranking." from http://rankabrand.de/sport-outdoormode/Vaude.

<References>
Barnett, M. L., Jermier, J. M. and Lafferty, B. A. (2006) 'Corporate Reputation: The Definitional Landscape', *Corporate Reputation Review*, Vol.9, No.1, pp. 26-38.
—— and Pollock, T. G. (2012) 'Charting the Landscape of Corporate Reputation Research', in Barnett, M. L. and Pollock, T. G. (Ed.), *The Oxford Handbook of Corporate Reputation*, pp. 1-15, Oxford University Press.
Bauchmüller, M. (2014) Auf Kante genäht Süddeutsche Zeitung, München: 2
Beck, D. E. and C. C. Cowan (2007) 'Spiral Dynamics-Leadership, Werte und Wandel: Eine Landkarte für Business, Politik und Gesellschaft im 21', Jahrhundert. Bielefeld.
Bhattacharya, C. B. and S. Sen (2004) 'Doing Better by Doing Good: When, why, and how consumers respond to corporate social initiatives', *California Management Review*, Vol.47, No.1, pp. 9-24.
Clean Clothes Kampagne Austria (2012) 'Firmen Check 2012: Outdoor', Available at:
http://www.cleanclothes.at/left-menu/firmen-check-2012/ Accessed April 6th 2015
Crane, A., Palazzo, G., Spence, L. J. and Matten, D. (2014) 'Contesting the Value of "Creating Shared Value"', *California Management Review*, Vol. 56, No.2, pp. 130-153.
Dahlsrud, A. (2008) 'How corporate social responsibility is defined: an analysis of 37 definitions', *Corporate Social Responsibility and Environmental Management*, Vol. 15, Issue1, pp. 1-13.
Einwiller, S. (2014) 'Reputation und Image: Grundlagen, Einflussmöglichkeiten, Management', in A. Zerfass and M. Piwinger (Ed.), *Handbuch Unternehmenskommunikation*, pp. 371-391, Spinger Wiesbaden.
European Commission (2001) 'GREEN PAPER: Promoting a European Framework for Corporate Social Responsibility' COM (2001) 366 final, Office for Official Publications of the European Communities.
—— (2011) 'COMMUNICATION FROM THE COMMISSION TO THE EUROPEAN PARLIAMENT, THE COUNCIL, THE EUROPEAN ECONOMIC AND SOCIAL COMMITTEE AND THE COMMITTEE OF THE REGIONS. A renewed EU strategy 2011-14 for Corporate Social Responsibility' COM (2011) 681 final, Office for Official Publications of the European Communities.
Fleischer, A. (2014) 'Reputation und Wahrnehmung: Wie Unternehmensreputation entsteht und wie sie sich beeinflussen lässt', Springer Wiesbaden.

Fombrun, C. J. (2012) 'The Building Blocks of Corporate Reputation: Definitions, Antecedents, Consequences', in Barnett, M. L. and Pollock, T. G. (Ed.), *The Oxford Handbook of Corporate Reputation*, pp.94–113, Oxford University Press.

—— and Van Riel, C. (1997) 'The Reputational Landscape', *Corporate Reputation Review*, Vol.1, No.1, pp. 5–13.

Frankental, P. (2001) 'Corporate social responsibility-a PR invention?', *Corporate Communications: An International Journal*, Vol. 6, Issue1, pp. 18–23.

Freeman, R. E., Harrison, J. S., Wicks, A. C., Parmar, B. L. and De Colle, S. (2010) *Stakeholder Theory: The State of the Art*, Cambridge University Press.

Greenpeace (2015) "Greenpeace Detox Fashion Campaign", Available at: http://www.greenpeace.org/international/en/campaigns/detox/fashion/ Accessed April 6th 2015

Hiß, S. (2006) *Warum übernehmen Unternehmen gesellschaftliche Verantwortung? Ein soziologischer Erklärungsversuch*, Campus Verlag.

ISO (2010) ISO 26000:2010, Guidance on Social Responsibility, International Organization for Standardization, Geneva.

Janney, J. J. and S. Gove (2011) 'Reputation ans Corporate Social Responsibility. Aberrations, Trends and Hypocricy: Reaktions on Forms Choioces in the Stock Option Bachdating Skandal', *Journal of Management Studies*, Vol. 48, Issue7, pp.1562–1585.

Kennedy, M. T., Chok, J. I. and Liu, J. (2012) 'What Does it Mean to Be Green? The Emergence of New Criteria for Assessing Corporate Reputation', in Barnett, M. L. and Pollock, T. G. (Ed.), *The Oxford Handbook of Corporate Reputation*, pp. 69–93, Oxford University Press.

Kiron, D., Kruschwitz, N., Haanaes, K., Reeves, M. and Goh, E. (2013) 'The Innovation Bottom Line-Finding from the 2012 Sustainability & Innovation Global Executive Study and Research Report', *MIT Sloan Management Review*, Vol. 54, Issue.2, pp. 69–73.

Loew, T. and Rohde, F. (2013) 'CSR und Nachhaltigkeitsmanagement. Definitionen, Ansätze und organisatorische Umsetzung im Unternehmen', Institute for Sustainability, Berlin.

McWilliams, A. and Siegel, D. (2001) 'Corporate Social Responsibilty: A Theory of the Firm Perspective', *Academy of Management Review*, Vol. 26, No.1, pp. 117–127.

Minor, D. and Morgan, J. (2011) 'CSR as Reputation Insurance: Primum Non Nocere', California Management Review Vol. 53, No.3, pp. 40–59.

Partnership for Sustainable Textiles (2015) Available at: http://www.textilbuendnis.com/index.php/en/, Accessed April 6th 2015.

Porter M. E. and Kramer, M. R. (2006) 'Strategy & Society: The Link between Competitive Advantage and Corporate Social Responsibility', *Harvard Business Review*, Vol. 84, Issue 12, pp.78–92.

—— (2011) 'Creating Shared Value', *Harvard Business Review*, Vol. 89, Issue1/2, pp. 62–77.

Rank a Brand (2015) '2015 Ranking.' Available at:

http://rankabrand.de/sport-outdoormode/Vaude, Accessed April 6th 2015.

Renner, M. (2011) 'Generating Trust via Corporate Reputation', Berlin.

Global RepTrack (2015) Available at:

http://www.reputationinstitute.com/research/Global-RepTrak-100, Accessed April 6th 2015

Schwaiger, M. (2004) 'Components and Parameters of Corporate Reputation-An Empirical Study', *Schmalenbachs Business Review*, Vol.56, pp. 46-71.

Schwalbach, J. and A. Schwerk (2014) 'Corporate Governance und Corporate Social Responsibility: Grundlagen und Konsequenzen für die Kommunikation', in Zerfaß, A. and Piwinger, M. (Ed.), *Handbuch Unternehmenskommunikation*, pp. 203-218, Springer-Verlag, Wiesbaden.

UN Global Compact and Accenture (2013) The UN Clobal Compact-Accenture CEO Study on Sustainability 2013, UN Global Compact.

Van Marrewijk, M. (2003) 'Concepts and Definitions of CSR and Corporate Sustainability: Between Agency and Communion', *Journal of Business Ethics*,Vol. 44, Issue2-3, pp. 95-105.

Walker, K. (2010) 'A Systematic Review of the Corporate Reputation Literature: Definition, Measurement, and Theory', *Corporate Reputation Review*, Vol.12, No.4, pp. 357-387.

II 企業ケース /Cases of "CSR and Strategy"

1 持続可能性と積水ハウスの戦略

和田　勇

2 企業と社会の相互作用としてのステークホルダー・レビュー による CSR の経営への統合化促進
　　―NEC のケース　　　　　　　鈴木　均・遠藤直見

3 BtoB 企業の CSR とブランド価値向上
　　―凸版印刷を事例として　　　　　　今津秀紀

4 キリンの CSV の取り組みについて
　　―「ブランドを基軸とした経営」による社会と企業の持続的な発展に向けて　　　太田　健・四居美穂子

1　持続可能性と積水ハウスの戦略

和田　勇

積水ハウス株式会社代表取締役会長兼 CEO

【要旨】

当社では「住宅は社会課題の中心に位置する」と考えている。また，2005 年に「サステナブル・ビジョン」を発表し，さまざまなステークホルダーに「環境価値」，「経済価値」，「社会価値」，「住まい手価値」を提供すべく事業活動を継続している。2013 年から，住宅を通じて社会課題を解決し，ステークホルダーとの共有価値を創造する 5 つの柱を「CSV 戦略」として位置付け，取り組みを強化している。当事例報告では当社が CSV 経営に取り組む背景，歴史，創出している価値等について記述する。

キーワード：価値創造，持続可能性，CSV 戦略，住宅は社会課題の中心，ネット・ゼロ・エネルギー，統合報告，マテリアル・アスペクツ

1. 住宅は社会課題を解決する鍵

住宅は社会や経済にさまざまな側面で大きなインパクトを与える事業であり，現在の社会課題を解決に導くことができる多くの可能性を備えている。

住宅に求められる最も基本的な役割は「家族の生命と財産を守るシェルター」というものだが，その他にも「家族の安らぎの場」や「健康な暮らしのベース」，さらには「子どもの教育の場」としても重要な役割を担っている。また，住宅が集まるコミュニティは人々の交流の場となり，地域の安全や文化を生み出す土壌となる。2011 年の東日本大震災から 4 年が経過した今も仮設住宅で不安な生活を余儀なくされている被災者の方々は多く，いかに住まいが

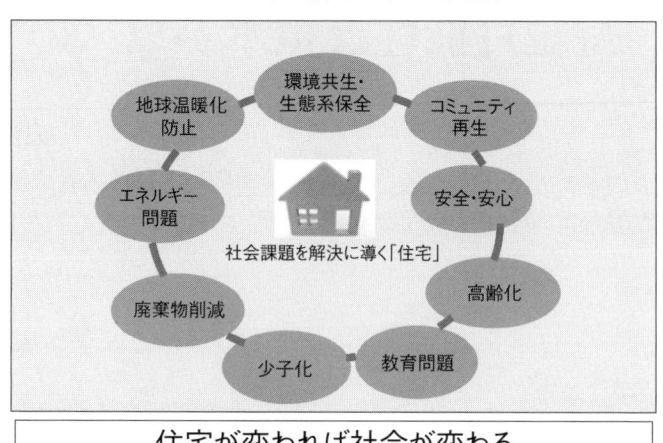

図1　住宅は社会課題の中心に位置する

(図：中央に「社会課題を解決に導く『住宅』」、周囲に「環境共生・生態系保全」「コミュニティ再生」「安全・安心」「高齢化」「教育問題」「少子化」「廃棄物削減」「エネルギー問題」「地球温暖化防止」)

住宅が変われば社会が変わる

心のよりどころとして大切か，地域のつながりが欠かせないものであるかが広く認識されることとなった。

近年の社会課題となっている環境・エネルギー問題や，近隣関係の希薄化によるコミュニティの崩壊，少子高齢化による世代間交流の減少などの問題も，住宅やまちのあり方と密接に関係していることから，住宅を通して社会に良い変化を生み出すことができると考えている。言い換えれば「住宅が変われば社会が変わる」といえるのではないか。年間約5万戸，累積にして223万戸という多くの住宅を提供している私たちだからこそ，社会課題の解決に向けて先陣を切って責任を果たしていかなければならないと考えている。

2. 積水ハウスが考える価値創造

当社は2005年，目指すべき「持続可能な社会」をビジョンとして定義し，それを実現・検証するために「経済」，「環境」，「社会」，「住まい手」の「4つの価値」によるバランスのとれた経営の実践を宣言した。また2006年には「4つの価値」を掘り下げた「13の指針」を定めた。この「サステナブル・ビジョン」が当社のすべてのCSR活動のよりどころとなっている。住まいづく

図2 積水ハウスが考える価値創造

「サステナブル・ビジョン」に基づく「4つの価値」と「13の指針」

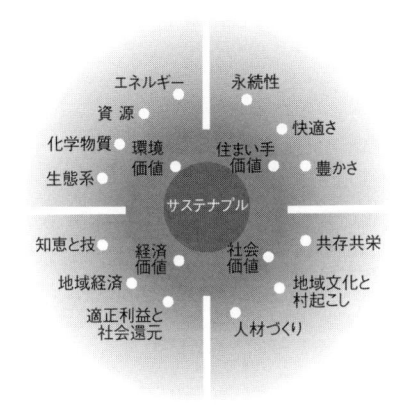

[図: 中央に「サステナブル」を置き、「環境価値」（エネルギー、資源、化学物質、生態系）、「住まい手価値」（永続性、快適さ、豊かさ）、「経済価値」（知恵と技、地域経済、適正利益と社会還元）、「社会価値」（共存共栄、地域文化と村起こし、人材づくり）の4領域を配置した円形図]

社会（お客様をはじめとしたステークホルダー）に価値を提供し、自社の競争優位に結びつける = Creating Shared Value

りという事業活動を通じて社会（お客様をはじめとしたさまざまなステークホルダー）に価値を提供し、自社の競争優位に結びつける CSV (Creating Shared Value) 戦略が当社 CSR 活動の中核となっている。

3. 「未来責任」と CSV 経営への歩み

当社は 1999 年に発表した「環境未来計画」を契機として、全社横断的な環境活動をスタートさせた。それまでも、高断熱・省エネルギー住宅の開発・販売は行っていたが、それは個別の取り組みであり、私たちの決断は「すべてにおいて環境を優先させる」ということであった。積水ハウスは「環境価値」に対する幅広い視野、多面的な新しい発想を持つことで、独自の取り組みを積み重ね、時代をリードするさまざまな技術・商品を生み出してきた。

そうした思いの根底にあったのが「未来責任」という言葉である。住宅は売れば終わりという商品ではない。住宅メーカーは、そこから始まるお客様の生

涯に責任を持つ必要がある。さらにその考え方を突き詰めていくと，住環境だけではなく，大きくは50年，100年先の地球環境にも責任を持つ必要があるのではないか。それが環境に対する「未来責任」である。今振り返ってみると，この環境への取り組みは「住宅が変われば社会が変わる」という私の主張やCSV経営と重なりあい，その原点になっていると考える。

商品開発においては，2009年に環境配慮型住宅「グリーンファースト」を経営戦略の軸として発売。グリーンファースト戦略は，2011年の世界初3電池（太陽電池，燃料電池，蓄電池）連動のスマートハウス「グリーンファーストハイブリッド」，さらには2013年に政府の施策でもあるネット・ゼロ・エネルギー・ハウスを先取りした「グリーンファースト ゼロ」に発展していく。

この環境技術を前面に押し出し，海外進出も果たした。各国で訴えたのは「現地の人たちに歓迎され，喜ばれ，価値を残す仕事がしたい」ということであった。国際的なCSV経営の実践である。その結果，業績も順調に推移している。環境技術は，クールジャパンの一つとして，日本の成長を支える大きな可能性を秘めていると確信している。

図3　持続可能性を経営の基軸に

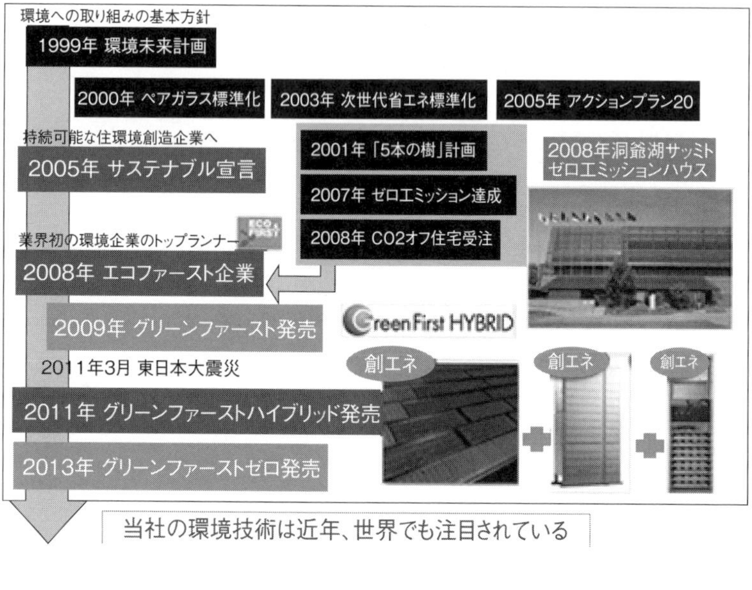

4. 積水ハウスグループの CSV 戦略

　積水ハウスグループは，住まいづくりを通じて「サステナブル・ビジョン」に基づき，多様なステークホルダーに対して価値を創造・共有することにより社会課題の解決に取り組んできた。
　CSV の一連の流れを IIRC の国際統合報告フレームワークのオクトパス・モデルをもとに図示すると図4のようになる。
　2013年に重点的に取り組む5つのテーマを選定し，「CSV 戦略」として位置付け，活動のさらなるレベルアップを図っている。
　テーマの選定にあたってはお客様，サプライヤー，投資家などのステークホルダーとのダイアログを経て GRI ガイドライン・G4 の特定標準開示項目の46側面について優先順位付けを行い，最終的に30の側面をマテリアル・アスペ

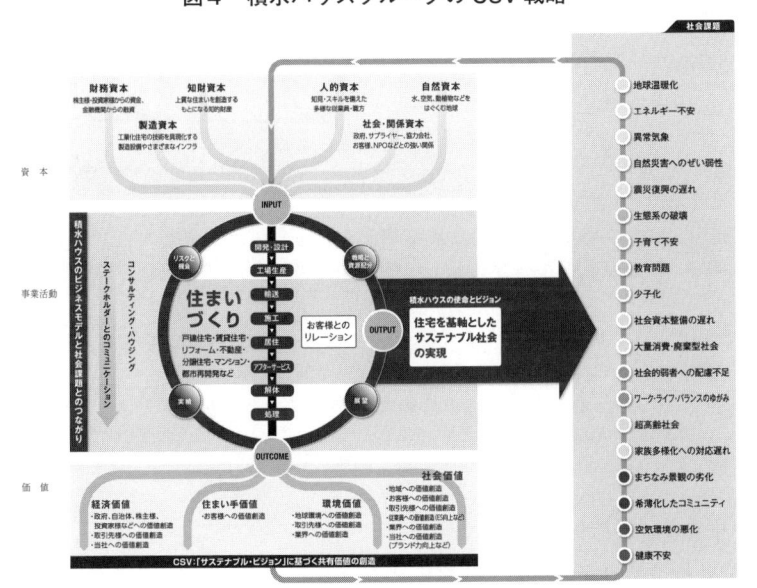

図4　積水ハウスグループの CSV 戦略

クツとして特定した。このアスペクツを具体的なCSR活動レベルに束ね，5つの「CSV戦略」を定めたのである。以下，それぞれの戦略について具体的に記述する。

4-1. 住宅のネット・ゼロ・エネルギー化

日本における全消費電力量のうち家庭部門が約3割を占めており，この削減に向けて住宅のゼロエネルギー化や，スマートハウスを中心とする電力需給の最適化，水素社会を見据えたインフラづくりなどが大きな課題となっている。

住宅メーカーとして，ネット・ゼロ・エネルギー・ハウス（ZEH）の普及に貢献するとともに，新エネルギーを利用した新しい暮らし方の実現にも挑戦していかなければならない。当社では，2013年4月にZEHを先取りした「グリーンファースト ゼロ」を発売し，2014年度の販売比率を59％まで高められたことは大きな成果だと考えている。2015年度は65％を目指し，取り組みを一層推進していきたい。

さらに，お客様が楽しみながら省エネできるよう，対話型ホームエネルギーマネジメントシステム（HEMS）「あなたを楽しませ隊」を開発した。今回開発した対話型HEMSは，画面に登場するキャラクターと対話しながら利用することができる仕組みである。また，お客様の情報を一元管理するため，お客様ごとに適した情報やサービスを提供することができる。

―――――――――――――――――――――――――――――
［持続可能性と共有価値1］
・政府が2020年までに標準的な新築住宅とする「ネット・ゼロ・エネルギー・ハウス（ZEH）」を積極推進し，地球温暖化・エネルギー問題の解決に寄与する。
・お客様にはより一層安全・安心・快適な住まいを提供し，他社との競争優位を確保する。
・当社にとっては受注量，受注単価の増大が期待できる。
―――――――――――――――――――――――――――――

4-2. 生物多様性の保全

生物多様性の恵みは，人々の暮らしや企業の事業活動の基礎となることを認

識し，早くから生物多様性の保全を重要なテーマと位置付け，取り組みを進めてきた。自然生態系の再生能力を超えない範囲で資源を利用するとともに，自然の循環と多様性を守るための配慮を行っている。その柱となるのが，住宅の原材料となる木材の持続可能な調達と，造園の際に地域の生態系に配慮する「5本の樹」計画である。2001年にスタートした「5本の樹」計画は多くのお客様のご理解と参加によって，2013年度には「5本の樹」計画の累計植栽本数が1000万本を超えることとなった。

また，2007年に独自の「木材調達ガイドライン」を策定し，生物多様性への配慮や，働く人の人権や労働慣行も含めた，持続可能な森林経営による木材調達方針を明確にした。木材の長く複雑なサプライチェーンの現状を把握するためには，サプライヤーの理解と協力が欠かせない。きめ細かいサポートを行いながら木材の持続可能な調達を推進していきたい。

［持続可能性と共有価値2］
・事業活動を通じ，第二の環境問題といわれる生物多様性への負荷を軽減する。
・お客様へも生物多様性の重要性を啓発する。
・当社にとっては外構・造園の受注・売上増大につながる。

4-3. 生産・施工品質の維持・向上

高品質な住宅を提供するためには，部材が高品質であることに加え，施工する職方の優れた技術が必要となる。しかし近年，職方の不足・高齢化や新規入職者の確保・育成が社会課題となっている。当社では，優れた職方の育成や，部材の自社生産を拡大することで，高品質の住宅を安定的に提供することを目指している。

当社が高い施工品質を維持できるのは，これまでの協力工事店との強いきずながあるからである。閑散期でも，協力工事店に長期にわたり安定的に発注し，職方の生活を支えることに注力してきたため，当社の仕事にロイヤルティーを持ち，質の高い施工をしてくれている。建築現場でのゼロエミッションに取り組んだ時には分別作業に積極的な協力が得られたし，東北地方での仮設住宅・復興住宅建設の際にも全国から多くの職方が応援に駆け付けてくれた。

東日本大震災以降，全社を挙げて復興支援を行ってきたが，2013年9月，木造住宅「シャーウッド」の陶版外壁「ベルバーン」の製造ラインを東北工場に新設した。これにより新たに100人の雇用を生み出すことができた。

[持続可能性と共有価値3]
・生産力・施工力の強化により品質向上・お客様満足向上を実現。
・協力工事店には安定経営を支援する。
・当社にとっては震災など有事の際に，臨機応変な対応をお願いできるとともに，労務費の高騰を抑止することができる。

4-4. 住宅の長寿命化とアフターサポートの充実

　日本の住宅は，平均寿命が約30年と欧米諸国に比べて著しく短いという特徴がある。住宅の資産価値が正しく評価されず，20年ほどで建物の評価がゼロに等しくなるのである。建てては壊すを繰り返していては，資源もエネルギーも大きな損失となる。このような状況では，愛着のあるまちなみや地域の文化がはぐくまれないし，国民の住居費負担も重くなり，心豊かな生活を送ることは難しいであろう。

　当社は，高品質・高耐久の住宅で，家族構成や住まい方の変化にも容易に対応でき，住まい手の愛着を生み出す工夫を凝らした住宅を提供している。さらにアフターサポートを充実させ，住まいを長期にわたる優良な社会資本とすることに取り組んでいる。

[持続可能性と共有価値4]
・多世代にわたって末永く付き合える住宅の提供を通じて良質な社会資産を形成
・耐久性の向上と快適で豊かな暮らしの追求を通じて，多様な形態の家族の暮らしを支援
・優良ストック住宅流通の活性化により，住宅投資額の逓減を阻止
・トータルなアフターサポートにより当社はストック収益を享受

4-5．ダイバーシティの推進

　少子高齢化が進み，労働力も減少する中，多様な人々の能力を活用していくことは，日本の活力を維持していくための鍵となる。女性も高齢者も，障がい者も外国人も，多様な人々が活躍できる社会に変えていかなければならないし，そうすることで変化に対し，しなやかに対応できる豊かな社会になるはずである。

　当社にとっても，人材のダイバーシティは不可欠なものと位置付けており，2014年2月には「ダイバーシティ推進室」を新たに設置した。多様な人材が創造性，革新性を発揮する組織のもと，社員がいきいきとした社会生活を送れるようにし，共通の目標に向かって共に取り組むことで力を結集し，事業を通じたイノベーションの実現を目指している。

　住宅は，暮らしと密接にかかわる仕事であるから，家事や育児を経験した方，暮らしの中で不便を感じている方の感性が生きる場がたくさんある。こうした事業上の意義も認識して，さらにダイバーシティを力強く推進していきたいと考えている。

［持続可能性と共有価値5］
・多様な人材の活躍の場を創出
・就労人口の確保に寄与
・当社にとっては多様な顧客ニーズへの対応，顧客目線での商品開発・提案，企業活動における様々な化学反応等が期待できる。

---

2 企業と社会の相互作用としてのステークホルダー・レビューによるCSRの経営への統合化促進
――NECのケース

鈴木　均
株式会社国際社会経済研究所代表取締役社長

遠藤直見
株式会社国際社会経済研究所グローバル・ビジネス・リサーチ部主幹研究員

【要旨】
NECは2004年度から，CSRレポートの作成プロセスを活用し，CSR経営の実績と中期・次年度の目標などに対するステークホルダー・レビューを実施している。ステークホルダー・レビューとは，企業活動に社会の視点（評価，期待など）を取り入れ，持続可能な経営に向けた改善・変革を促すツールである。本稿では，社会的責任の国際規格であるISO 26000に基づき，NPO法人「CSRレビューフォーラム」（NPO／NGO，労働専門家，消費者専門家，ISO 26000策定経験者などから構成）との協働（エンゲージメント）により実施しているステークホルダー・レビューが，CSRコンセプトの経営戦略や事業活動への統合化に有効なツールとして機能したことが確認できたので，その概要，成果と課題，獲得した知見などを報告する。

キーワード：CSR，ステークホルダー・レビュー，ステークホルダー・エンゲージメント，CSRの経営への統合，ISO 26000，持続的な成長，中期経営計画，デューデリジェンス，ビジネスと人権，CSR調達

1. NECのCSR経営

　NECでは，CSR経営を持続可能な経営と捉え，NECグループビジョン2017の掲げる「人と社会にやさしい情報社会をイノベーションで実現するグローバルリーディングカンパニー」を目指している。そのCSR経営は以下の3つの方針に基づき推進される。
　　(1) リスク管理・コンプライアンスの徹底
　　(2) 事業活動を通した社会的課題解決への貢献
　　(3) ステークホルダー・コミュニケーションの推進
　(3)には，「アニュアル・レポート，CSRレポートなどによるステークホルダーへの情報開示と説明責任」および「ステークホルダー・エンゲージメント（ステークホルダー・レビュー）に基づく経営へのCSRの統合」の2つが含まれる。本稿では，後者に焦点を絞り，2004年度からの10年に渡るNECの取り組みについて述べる。

2. ステークホルダー・レビューによる経営へのCSRの統合化

2-1. NECのステークホルダー・レビュー

　NECでは，CSR経営の一環として，2004年度から，社会課題を熟知したNPOとの対話とレビュー（ステークホルダー・レビュー）に継続的に取り組んできた。ステークホルダー・レビューでは，コーポレートスタッフなどの主管部門（人事，総務，法務，内部統制，調達，経営システム，環境，CS，品質，生産など）が，NPOなど社会セクターとの対話をとおして，自らの取り組みに「社会の声」（NECへの評価，提言など）を取り入れ，改善に繋げている。これは，2015年6月に運用が開始されたコーポレートガバナンス・コードの原則2-3「社会・環境問題をはじめとするサステナビリティ（持続可能性）を巡る課題」にも対応している。

　これまでのステークホルダー・レビューの経緯は，図1に示されている通り，

図1　ステークホルダー・レビューの経営への統合

- 目的：持続可能な経営の改善,説明責任と透明性確保
- 特徴：CSRレポートを「てこ」にしたNPOとのステークホルダー・エンゲージメントに基づくPDCAサイクル推進

ステージ1：2004年度〜
取り組みの改善
・レビュアー：NPO（IIHOE）
・レビュイー：実務担当責任者など

ステージ2：2011年度〜
ISO 26000に基づく改善
・レビュアー：NPO（CSRレビューフォーラム）
・レビュイー：CSR担当役員および各主管部門長など

ステージ3：2014年度〜
経営戦略への統合化と重点領域へのフォーカス
・レビュアー：NPO（CSRレビューフォーラム）
・レビュイー：副社長兼CSO、経営参画部門も参画

ISO 26000:社会的責任の国際規格 2010年11月発行
CSO:Chief Stratagic Officer

縦軸：経営へのCSRの統合レベル
横軸：2004年／2011年／2014年

3段階に分けられる

・ステージ1（2004〜2010年度）

　レビュアーをNPO「IIHOE」（人と組織と地球のための国際研究所），レビュイーを人事や調達などの各主管部門の主に実務担当責任者とし，課題の明確化と気づきを抽出し，取り組みの改善につなげていった。

・ステージ2（2011年〜2013年度）

　レビュアーをNPO「CSRレビューフォーラム（CRF）」，レビュイーをCSR担当役員とCSR担当部門，各主管部門の部門長および実務担当責任者とした。レビューは，社会的責任のグローバルスタンダードであるISO 26000に基づき実施した。これにより，自部門の取り組みの課題とその取り組みが国際基準と比較してどのくらいのレベルにあるかの把握が進んだ。

・ステージ3（2014年度〜）

　ISO 26000に基づくCRFとのレビューを継続。2014年度はガバナンスを中心とする従来のレビュー領域の中から特に重点課題として確認された人権，CSR調達（公正な事業慣行），環境の3つに加え，経営戦略（2015中期経営計

写真1 ステークホルダー・レビューの模様

画）の社会視点でのレビューを実施した。このレビューには，経営戦略と CSR を担当する役員の副社長，経営企画本部も参画した。

2-2. NPO 法人「CSR レビューフォーラム」（略称：CRF）

CRF とは，持続可能な社会作りに取り組む NPO（民間非営利組織）であり，人権関係等の NGO，労働や消費者関係の専門家，CSR に関する有識者，ISO 26000 策定に参加した専門家などが共同設立した中間組織であり，CSR 各領域での知見と ISO 26000 をベースに企業活動への第三者レビューを行う「CSR レビュープログラム」を提供している。

2-3. ISO 26000 に基づくステークホルダー・エンゲージメント

ISO 26000 は要求事項を含まないガイダンス文書であり，「認証，規制，契約のための使用を意図しない」，「マネジメントシステム規格ではない」などの特徴を持つ。そのため，規格内容の実効性はステークホルダー・エンゲージメントで確保することを重視している。ISO 26000 が定めるステークホルダー・エンゲージメントの形態は，「組織とひとり（または一組）以上のステークホルダーの対話」であり，「ステークホルダーの意見を聴く機会を設けること」を目的としている。また，その本質的な特徴は，「組織からの情報提供を含む双方向のコミュニケーションを必要とすること」であり，「相互に有益な目的達成のためのパートナーシップ」である。

また，ISO 26000 が定めるステークホルダー・エンゲージメントの企業への

メリットとしては，「情報発信内容に関する透明性向上に有効であること」，「報告内容の検証を行い，その結果の証明によって信頼性が向上すること」，「ステークホルダーの意見に留意した意思決定に有効であること」，「組織のパフォーマンスの確認と改善に寄与すること」などが挙げられる。

2-4. ISO 26000 に基づくレビューの視点

ステークホルダーによるレビューの実効性を高めるため，主に以下の4つの点を重視した。

(1) グローバル標準との比較，すなわち自社の取り組みが，ISO 26000 が求めるグローバルなレベルに達しているかどうかを確認する。
(2) 持続可能な経営に必要な要件の網羅性，すなわち自社の取り組みが ISO 26000 の7中核主題，36課題を満足しているかどうかを確認する。
(3) 社会的責任の7原則を重視，すなわち自社の取り組みが，以下の7原則を重視しているかを確認する。
　　①説明責任，②透明性，③倫理的な行動，④ステークホルダーの利害の尊重，⑤法の支配の尊重，⑥国際行動規範の尊重，⑦人権の尊重
(4) 経営への CSR の統合度合い（深さ，広がり）
　　「深さ」とは，ステークホルダー・エンゲージメントやデューデリジェンス（企業活動が社会に及ぼす負の影響の特定，回避，軽減などの活動）の取り組みが企業活動の中にしっかり組み込まれ，PDCA が回っているかどうかを確認する。「広がり」とは，企業活動の影響力と責任の範囲に関してである。例えば，マネジメントの範囲が国内グループ会社までか，海外グループ会社までか，あるいはバリューチェーン全体まで包含しているかどうかを確認する。

2-5. ステークホルダー・レビューのプロセス（概要）

ステークホルダー・レビューのプロセスは，「事前準備」，「対話とレビュー」，「課題の抽出」，「改善計画の策定」，「情報開示」および「継続的改善」の各ス

2 　企業と社会の相互作用としてのステークホルダー・レビューによるCSRの経営への統合化促進　　173

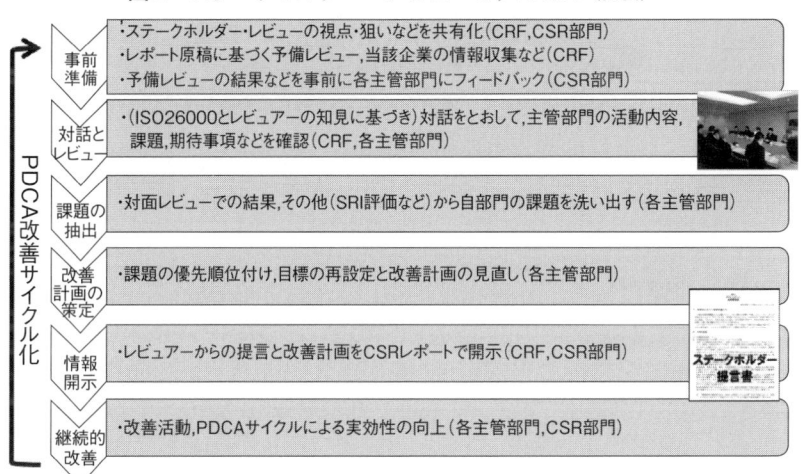

図2　ステークホルダー・レビューのプロセス（概要）

PDCA改善サイクル化

- **事前準備**
  - ステークホルダー・レビューの視点・狙いなどを共有化（CRF,CSR部門）
  - レポート原稿に基づく予備レビュー,当該企業の情報収集など（CRF）
  - 予備レビューの結果などを事前に各主管部門にフィードバック（CSR部門）
- **対話とレビュー**
  - （ISO26000とレビューアーの知見に基づき）対話をとおして,主管部門の活動内容,課題,期待事項などを確認（CRF,各主管部門）
- **課題の抽出**
  - 対面レビューでの結果,その他（SRI評価など）から自部門の課題を洗い出す（各主管部門）
- **改善計画の策定**
  - 課題の優先順位付け,目標の再設定と改善計画の見直し（各主管部門）
- **情報開示**
  - レビューアーからの提言と改善計画をCSRレポートで開示（CRF,CSR部門）
- **継続的改善**
  - 改善活動,PDCAサイクルによる実効性の向上（各主管部門,CSR部門）

テップから構成される。詳細は，図2に示す通りである。なお，括弧内は担当組織を示している。

「対話とレビュー」が実際に企業とNPOが対話する機会であるが，その後の「課題の抽出」，「改善計画の策定」が重要になる。「対話とレビュー」で得られた情報，その他（SRI／メディア評価など）の情報などから自部門の課題を洗い出し，優先順位付け，目標の再設定と改善計画（施策含む）の見直しを実施する。これらのプロセスを丁寧に実施することがステークホルダー・レビューの実効性の向上に繋がる。

2-6. レビューアーへの要望・期待事項

実効的なレビュー実施のために，レビューアーには「第三者でチームを構成すること」，「企業および業界の知識を保有すること」，「価値向上（攻めのCSR）視点が必要であること」，「監査的な手法ではなく，コンサル的なアプローチで臨むこと」を要望し，それらを考慮いただいたことが実効性につながった。詳細は図3に示している通りである。

図3　レビュアーへの要望・期待事項

第三者でチーム構成	・7中核主題に精通したNPO（NGO，労働関係，消費者関係，識者など）による社外の専門家集団
企業，業界の知識	・当該企業や業界についてある程度の知識を保有（ICT分野，BtoBビジネスの特徴など）
攻めのCSR視点	・ISO 26000はコンプライアンス基調（守り）。企業価値向上に繋がる攻めの視点も必要（ダイバーシティ，消費者課題，コミュニティ参画など）
監査的な手法はNG	・監査的ではなく，コンサルタント的なアプローチ（対話重視）で臨む ・「独立性，中立性」VS.「企業の立場への理解」のバランス

3. ステークホルダー・レビューによる成果

ここでは，2011年度以降（ステージ2，3）のCRFとのステークホルダー・レビューに焦点を当てて，その主な成果について考察する。

3-1. 概要

ステークホルダー・レビューを継続することによって得た成果は主に以下の5点に集約される。1点目は，社会が企業に求める期待やニーズ，社会課題をステークホルダーとの対話をとおしてより深く理解することができたこと。2点目は，ステークホルダーからの評価に基づき課題の明確化と改善推進ができたこと。3点目は，国際標準との比較による自社の取り組みのレベル感が把握できたこと。4点目は，関係部門の当事者意識の高まりとCSR部門との連携が進展したこと。5点目は，ステークホルダーに対する信頼性，透明性や説明責任が進んだことである。

以下にもう少し具体的に述べる。

(1) 2011～2013年度（ステージ2）：ISO 26000に基づく取り組みの改善。
　　ステークホルダー・レビューに対する各主管部門・責任者の理解と協力度合いが向上し，各主管部門でのPDCAサイクルが徐々に定着してきた。この結果，ISO 26000の7中核主題を切り口とするNEC

のCSRへの取り組みへの網羅的な課題抽出と改善が進捗した。
(2) 2014年度〜（ステージ3）：経営戦略への統合化，重点領域へのフォーカス。

経営戦略（2015中期経営計画）の社会視点でのレビューに着手した。レビューの視点は，社会課題の捉え方，中長期の価値創造ストーリー，価値評価KPI（Key Performance Indicators）などであり，副社長兼CSO（Chief Strategic Officer：戦略担当役員），経営企画部門なども参画した。この結果，社会・環境問題をはじめとするサステナビリティを巡る課題への全社的な議論が進展した。また，人権，CSR調達（公正な事業慣行），環境などは過去のステークホルダー・レビューで確認された重点領域であり，継続してレビューした。

3-2. 領域別

ステークホルダー・レビューによる成果を，CSR経営全般，人権，CSR調達（公正な事業慣行）について以下に示す。

3-2-1. CSR経営全般

NECグループの重要課題を「デューデリジェンス」（自社グループの影響力と責任の範囲を考え，自社の企業活動が社会に及ぼす負の影響を把握・回避・軽減する取り組み）と「本業でのCSR推進」，すなわち社会の持続可能な発展と競争優位の両方に貢献する価値創造の取り組みと捉え，継続的にレビューを実施した。この価値創造の取り組みはPorter and Kramer（2011）のCSV（Creating Shared Value）の考えに近い。前者の課題については，先ずは「人権デューデリジェンス」に焦点を絞った。これについては，(2) 人権で詳細を述べる。後者の課題については，2013年4月に「2015中期経営計画」が策定・公表されたことにより，「社会価値創造型企業への変革」という会社の目指す方向性と実質的に統合された。この機会を捉え，2014年度のステークホルダー・レビューでは，経営戦略担当役員（CSR担当も兼務）と経営企画本部も参画し，NECの持続可能な成長に向けた「社会の声（NECへの評価・期待）」を傾聴する機会を設定した。

表 1　ステークホルダー・レビューによる成果：CSR 経営全般

	CRF のコメント　◎は改善事項
2011 年度	・NEC グループの最重要課題は「組織全体への社会的責任の統合」 ・具体的には以下 2 点。 　①デューデリジェンス：自社グループの影響力と責任の範囲を考え，自社の企業活動が社会に及ぼす負の影響を把握・回避・軽減する取り組み。 　②本業での CSR 推進：社会の持続可能な発展と競争優位の両方に貢献する攻めの取り組み（CSV：Creating Shared Value）。
2012 年度	◎ CSR 部門と各主管部門との連携が進展（今後は経営企画部や事業部門との連携進展に期待）。 ・デューデリジェンスの考え方の社内浸透と体制作りを期待。 ・多様性に関するビジョンを描き，短期と長期の目標策定を期待。
2013 年度	◎経営戦略推進基盤としての「多様性」の重要性について理解が進展。 ◎人権についてのデューデリジェンスに進展。 ・「社会価値創造」観点での事業評価指標による自己評価が必要。 ・経営企画部と CSR 部を中心に関係部署が協働して事業評価指標を作る。
2014 年度	◎ CSO（戦略担当役員），経営企画部を交えて社会視点での中計レビューを実施。 ・世界の主要な社会課題（人口爆発，気候変動等）の解決貢献をコミットする。 ・BtoB 事業にて顧客が解決しようとする社会課題を NEC はどう支援するかを示す。 ・社会価値創造の 2 つのアプローチに取り組む。 　①社会課題を切り出し，新事業（製品・サービス）を開拓。 　②従来事業を社会課題解決の観点で自己点検（バリューチェーン／事業プロセス見直し）。 ・社会価値の測定にチャレンジする─損益に加え，「社会課題をどれだけ解決したか」，「QOL がどれだけ向上したか」等の観点での事業評価指標の策定，試行に期待。

　表 1 に掲載されているような指摘事項については，現在，経営企画本部と CSR 部門が関連部門と連携しながら，組織横断的な改善検討に取り組んでいる。

3-2-2. 人権

　主な課題は，NEC グループでの人権デューデリジェンスの構築である。NEC では，当初から，人権問題は，グループ従業員および取引先の労働環境に及ぼす負の影響（長時間労働，職場での差別など）と NEC の製品・サービスおよび事業活動がお客様や地域社会の住民などに及ぼす負の影響（製品の安全問題，環境問題および ICT 企業の立場としての個人情報保護やプライバシー問題など）に大別されるという前提で取り組みを進めてきた。ここでは，仮に前者を

2 企業と社会の相互作用としてのステークホルダー・レビューによるCSRの経営への統合化促進　177

表2　ステークホルダー・レビューによる成果：人権

	CRFのコメント　◎は改善事項
2011年度	◎「NECグループ企業行動憲章」と「NECグループ行動規範」の中で、「人権の尊重」を打ち出している。 ・人権の認識が「差別的扱いの禁止」に留まっており、グローバルな人権概念の理解が課題。
2012年度	◎人権DDの基本的な仕組みづくりに着手。 ◎意識浸透面で国内の人権研修は充実している。 ・人権尊重について、NECが目指すグローバルな企業像、達成目標を設定し、段階的にステップを踏むことを期待。
2013年度	◎人事部とCSR部が連携して海外拠点での人権研修を推進している。 ・人権DDの全体像の整理とPDCA化のプランが必要。 ・人権DD構築では、事業部を含む横断的PJチームの編成が重要。 ・適切な外部ステークホルダーとの対話や連携を期待。
2014年度	◎CSR、人事、調達が連携して人権DDを構築している。 ・今後は以下を期待。 ①各国・地域の人権課題に対応したDD推進。 ②事業活動による人権影響を現地ステークホルダーの声も踏まえ整理。 ③現地NPOの取り組み状況と課題把握。 ④人権DDの主要ターゲット、成果と課題等の進捗開示（構築途中でのステークホルダーとの対話も）。

「（労働者への影響が発生する）労働人権」、後者を「（市民に影響が及ぶ）ビジネス人権」と呼ぶ。

「労働人権」については、既にCSR、人事、調達の各部門が連携し、粛々と取り組みを進めてきている。詳細は表2に示す通りである（表中に記述されているDDはデューデリジェンスを示す）。今後は、ビジネス人権におけるデューデリジェンス構築に取り組むことが期待されている。

3-2-3. CSR調達

主に、中国・アジアを中心とする海外サプライヤーの現地診断の展開と紛争鉱物問題対応の2点の課題がある。前者については、2012年頃から、中国の現地法人とも連携し、徐々に現地診断の仕組みを整備してきた。今後の課題は、診断プログラムの質・量の拡充および現地診断員の育成などである。また、紛争鉱物問題については、国内の業界団体JEITA（日本電子技術産業協会）が主宰する「責任ある鉱物調達検討会」の幹事会社として、業界連携活動の継続な

表3　ステークホルダー・レビューによる成果：CSR 調達

	CRFのコメント　◎は改善事項
2011年度	◎1次サプライヤーへの書面調査などを継続して実施している。 ・今後の課題は，実地確認，業界共同での取り組みなど。
2012年度	◎中国，インドを含む各拠点によるサプライヤーの実地調査（枠組み）の計画策定を評価する。 ◎紛争鉱物問題対応では，米国 SEC のルール化に備えて体制を整備している。
2013年度	◎CSR 部と調達部が連携し，実地診断プログラムの試行，現地ステークホルダーとの対話などを実施している。 ◎JEITA「責任ある鉱物調査検討会」に幹事会社として参画している。 ・2次以降のサプライチェーンマネジメント，アセッサー養成等が課題。
2014年度	◎紛争鉱物では JEITA 検討会にも参画し，可能な範囲での対応を実践していることを評価する。 ◎実地診断プログラムの体制は整いつつある。 ・実地診断プログラムにおいて，今後は以下を期待。 ①リスクの高いサプライヤーを選定し，プログラムの質・量を拡充。 ②2次，3次サプライヤーへの実地診断の実施。 ③外部ステークホルダーの実地診断への参画など。

ど可能な範囲での対応を実施してきている。

4. 獲得した知見

　ステークホルダー・レビューをとおして獲得した知見は以下の6点に集約される。

(1) 社会課題・ニーズ把握のための有効な手段

　ステークホルダー・レビューは，社会課題とそれに裏打ちされた社会ニーズ把握のための有効な手段の1つであることが確認できた。持続可能な経営のためには，収益確保と社会課題解決の両立が求められる。そのためには，自社内だけでなく，外部のステークホルダーの意見を取り入れ，それを「てこ」とした推進が必要となる。一般的に企業は，「課題をどう解決するか」という方法論（How）は得意だが，「当該企業が解決に取り組むべき社会課題は何か（What）」，「その社会課題がなぜ発生するのか（Why）」といった視点が弱く，その点を補完する意味でもステークホルダーとの対話が企業に気づき（インサ

イト）を与えてくれることが確認できた。
(2) レビューの継続性が重要
　ステークホルダー・レビューは，企業と社会の相互作用を促す触媒であり，継続的に実施することで企業とステークホルダーとの信頼関係が構築されることが確認できた。両者の信頼関係に基づき，PDCAサイクルを継続的に廻すことで改善が進展し，課題が絞り込まれる。その結果，ステークホルダー・レビューの費用対効果の向上にも繋がる。
(3) 経営トップの参画が鍵
　相互作用の効果を最大化させるには適切な参加者の選定が重要である。企業においては実務責任者の参画も必要だが，経営と組織へのCSRの統合を進めるには，経営トップ，特に経営・事業戦略機能を管掌する役員および関連部門の責任者の参画が鍵となることが確認できた。
(4) CSR部門のリードと主管部門の協力，トップのサポートが必須
　CSR部門が旗振り役となって人事，経営企画，調達，法務などの関係主管部門を巻き込んでいくことが成功の鍵であることが確認できた。そのためには，主管部門の理解と協力，それらを支える経営トップのサポートとコミットが欠かせない。
(5) レビュアーとしてのNPO・NGOの参画が不可欠
　レビュアーには，生活者・消費者，さらには社会的弱者などの代弁者であり，また社会課題解決に継続的に取り組んできているNPO・NGOを欠くことができない。NPO・NGO以外には，労働関係／消費者関係／CSR関係の有識者などの参画も必要であり，構成メンバーには真のステークホルダー（社会的弱者など）の利害の認識とその代弁者という立場の理解，社会課題に対する専門性などが求められる。
(6) 対話手法が中長期の改善には効果的
　ステークホルダー・レビューは監査ではなく，企業とステークホルダーが対等な立場で真摯に対話し，納得感を醸成しながら課題共有，改善に繋げていくものであり，監査的手法と比べ，中長期視点での改善に向いていることが改めて確認できた。

5. ステークホルダー・レビューの継続的な発展に向けて

　NECのステークホルダー・レビューは，ISO 26000をベースとしている。ISO 26000は組織（企業）に求められる社会的責任の要素や要件を網羅するなど優れた特徴を多く持つが，一方ステークホルダー・レビューの実践を通じていくつか課題があることも判明した。ステークホルダー・レビューの継続的な発展のために，それらへの対応について以下の4点に分けて述べる。

(1) 価値向上に繋がる視点を強化する

　ISO 26000の内容は組織が順守すべき事項を中心に記述されていて，いわゆる「守り」志向が強く，企業の価値向上に繋がるいわゆる「攻め」の部分に関する記述が少ないことがあらためて確認できた。経営戦略へのCSRの統合を深化させるには，企業の独自の工夫が必要である。例えば，中期経営計画の社会視点でのレビュー，また戦略担当役員や経営企画部門，事業部門の参画などである。

(2) 顧客と"顧客の先の社会や顧客"との関係を意識的に考える

　「消費者課題」に代表されるように，ISO 26000の内容はBtoCビジネスを主対象としている傾向がみられる。BtoBビジネスでは，BtoBtoCとして，顧客と"顧客の先の社会や顧客"との関係を意識的に考えることが必要である。例えば，消費者課題の項目に加え，コミュニティ参画と発展などが対象となる。

(3) 最近の重要課題を積極的に取り入れる

　最近の重要課題である事業継続に関するBCP（Business Continuity Plan）／BCM（Business Continuity Management）などについては自主的にレビュー項目に設定することが必要である。

(4) 7中核主題と企業内組織が非整合

　企業内の組織（主管部門）とISO 26000の7中核主題とは必ずしも1対1に対応が取られているわけではない。例えば，ISO 26000の6-2「人権」は，人事部だけの担務ではない。CSR，調達，法務，事業部門などを含む全社的な取り組みが必要になる。6-8「コミュニティへの参画と発展」には，社会貢献活

動だけでなく，ビジネス側面もある。また，社会貢献活動は他の章にも記述がある（6-5「環境」，6-7「消費者課題」など）。ステークホルダー・レビューを実効的に実施するためには，これらの非対称性をうまく整合させるなどCSR部門の企画・調整力が鍵となる。

何れにしても，ステークホルダー・レビューの継続的な発展のためには，主管となるCSR部門の創意工夫とリード，そして経営トップによる継続的かつ強力なサポートが必須であることは言うまでもない。

＜参考文献＞

ISO/SR国内委員会監修（2011）『ISO 26000：2010 社会的責任に関する手引き』，日本規格協会編．

Porter, M. E. and Kramer, M. R. (2011) 'Creating Shared Value', *Harvard Business Review*, Vol. 89, Issue1/2, pp.62-77.

3 BtoB企業のCSRとブランド価値向上
──凸版印刷を事例として

今津秀紀
凸版印刷株式会社 コーポレートコミュニケーションチーム チームリーダー

【要旨】
BtoB企業が常に抱えている悩みがある。それは，BtoB企業であるがゆえのブランド力の弱さである。BtoBなら顧客企業に理解してもらえれば十分ではないかと言う人も多い。確かに既存顧客との取引だけを考えるならばその通りだろう。しかしながら，新規顧客の開拓や優秀な人材の獲得などの成長戦略を総合的に進めるためにはブランド力が不可欠になる。

　CSRはBtoB企業のブランド形成に大きな影響を及ぼす。誰からどのような評価を獲得するかの組み立て次第で企業ブランドは大きく向上する。そこでBtoB企業がどのステークホルダーからどのような評価を獲得して行くべきなのかをコミュニケーションの実行施策から説明する。

キーワード：BtoB，CSRブランディング

1. はじめに

　日本でサステナビリティ元年やCSR元年と呼ばれた2003年から数えて10年以上が経過した。CSRに関する情報開示を行っている企業も1,000社を超えると言われてから久しい。特に大手企業ではCSR活動を年次のCSRレポートで公表することがもはや当たり前になっている。そして，これらの開示された情報とメディアや調査機関が行うアンケート調査などからさまざまな企業評価

や格付が行われている。

　CSR がグローバルに広がり始めた当初より CSR が最も影響を与えるのは企業のレピュテーションやブランドと言われてきた。そして，その影響は昔も今も変わらない。一般社会で CSR とブランドというと，消費者から直接影響を受ける BtoC 企業をイメージしがちだが，実は BtoB 企業も同様なのである。そこで，凸版印刷という典型的な BtoB 企業を事例にして CSR とブランドマネジメントがどのように行われているのかを CSR コミュニケーションを中心に紹介する。

2. 凸版印刷（以下トッパン）とはどのような会社なのか

2-1. トッパングループの事業概要

　トッパングループは，凸版印刷株式会社および関係会社 180 社（子会社 154 社，関連会社 26 社）で構成されている。グループ売上高は約 1 兆 5 千億円。従業員数約 4 万 8 千人。世界最大級の総合印刷会社である（データは 2014 年 9 月現在）。

　印刷技術を核にした 3 つの事業分野を展開している。1 つ目が情報コミュニケーション事業分野。この分野は世間一般にイメージする印刷に一番近い。書籍や雑誌などの印刷，企業のパンフレットやカタログ類，また，金融の製品ではキャッシュカードやクレジットカードも扱っている。近年はインターネットの普及により，Web サイトや映像，また，電子書籍などの IT を活用したコンテンツへと拡大している。2 つ目が生活環境事業分野。身近な製品には食品などのパッケージ類がある。菓子，飲料やお酒などの包装材，また，日用品や化粧品の包装材などがある。最後の 3 つ目がマテリアルソリューション事業分野。エレクトロニクスの部材や建築用の資材といえば少しは分かりやすいだろうか。液晶ディスプレイの部材，太陽光パネルの部材，半導体の部材，壁紙などの建築部材を生産している。

2-2. 日本独特の典型的な BtoB 企業

　これらの事業分野で構成されるトッパングループだが，顧客は消費者ではなく企業になるため，世間一般にはほとんど知られていない典型的な BtoB 企業である。規模は大きいので社名くらいは知られているだろうが，事業内容は直接取引のある企業担当者以外にはまず知られていない。また，トッパンのような総合印刷会社は日本独特の業態であり海外には同規模の総合印刷会社というものが存在しない。BtoB 企業，かつ日本独特の総合印刷会社であるがゆえに企業広報や企業ブランディングにはいつも苦労している。

3. トッパンの CSR の考え方

3-1. 環境方針のような CSR 方針は持っていない

　CSR 元年と呼ばれた 2003 年からの数年間で CSR 憲章や CSR 方針を発表する企業が登場してきた。日本企業の多くは環境の認証規格である ISO 14001 を取得している。これらの企業は環境マネジメントの一環で環境方針を策定し，公表している。当時の CSR 推進は環境部門が兼任していたケースが多く，同様の推進体制を進めたことから CSR 方針が策定された。2000 年代の後半には社会貢献活動方針を策定する企業も増えてきた。

　しかしながら，環境活動や社会貢献活動とは違い，CSR は企業活動全体に関わるものである。事業活動の一側面に限定された活動ではない。そこで，CSR 方針を新たに策定する代わりに，企業理念やビジョンを実践することがまさに CSR であるとした考え方に移って来たのである。

3-2. 企業像の実現

　トッパンは 1900 年に創業した。100 周年を迎えた 2000 年に，社会や地球環境と調和しながら成長を続けるための基本的な考え方や活動の方向性を「TOPPAN VISION 21」として定めた。これは「企業像」と「事業領域」からなり，企業像の実現に取り組むことが，社会とトッパンがお互いに持続的に発展していくことに貢献するという考え方である。

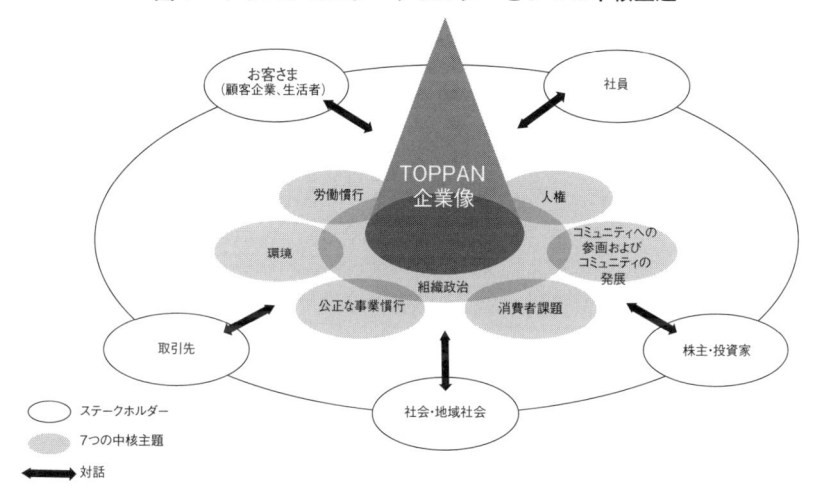

図1 トッパンのステークホルダーと7つの中核主題

3-3. ISO 26000 を参考に CSR 活動を推進

CSR 活動を推進するにあたっては，ISO 26000 をその活動の基盤にしている。後半の頁で触れるが，トッパンの顧客は企業であり，毎年，複数の国内海外企業から CSR に関する監査も受ける。そこで，グローバル基準で判断がしやすい ISO 26000 を積極的に取り入れて進めることにしたのである。

4. 社外からのトッパンに対する評価

4-1. CSR による企業評価のブーム

筆者は，国内企業に環境報告書が広がり始めた 1999 年以来，環境や CSR に関するコミュニケーションを担当している。振り返ってみると，CSR に関するさまざまな評価やランキングが発表されたのは，2000 年代後半の 2006 年〜2009 年あたりであったように思う。現在も CSR や環境に関する評価ランキングやアワードの発表は続いているが，すでにその使命を終えて無くなったものもあり，現在はブームを乗り越えてきたものと，ブームの後から新たに登場し

たものになっている。

4-2. トッパンが得た国内外からの評価

トッパンはまさにこのブームといえる 2006 年～2009 年にかけて高い評価を獲得することが出来た。特に高評価を受けたものを挙げると，

- 企業の社会的責任　世界 350 社ランキング第 10 位　NEWSWEEK 日本版（2008）
- 国内企業 CSR 活動ランキング第 1 位　日本財団（2007）
- 世間に自慢できる企業ベスト 100 第 1 位　読売ウィークリー（2007）
- CSR 報告書での情報開示が進んでいる企業ランキング第 1 位　フジサンケイビジネスアイ（2009）
- サステナビリティ報告書賞 最優秀賞 東洋経済新報社／グリーンリポーティングフォーラム（2008）
- 環境コミュニケーション大賞 持続可能性報告優秀賞 地球・人間環境フォーラム／環境省（2008）

また，最近のグローバルでの評価では次のようなものがある。

- Global Compact 100 Index（2014）
- FTSE4Good（2014）

世間に知られていない BtoB 企業が，BtoC の優良企業と肩を並べたわけである。そして，これらの高評価を受けた後は，CSR レポートの請求数や当社の CSR サイトへのアクセス数が大幅に増加した。

5. コーポレートブランドやマーケティングからみるトッパンの立ち位置

トッパンの考える CSR とブランドマネジメントを説明するために，企業としての立ち位置を先に紹介する。

・BtoB 企業であること

繰り返し説明してきたが，トッパンの顧客は一般生活者ではなく企業である。そのために自社の広告宣伝はほとんど行わない。そして，顧客企業の製品・

サービス用の部材や販促物を納めることから，トッパンのブランドマークが付いた製品やサービスもほとんどない。

・サプライヤーであること

　顧客企業へ部材や販促物を納めるということは，数千社の顧客企業のサプライヤーである。実際，CSR調達に関するアンケート調査や監査を受ける機会は多い。特に外資系の食品企業からは，国際的な認証機関による監査結果を求められることが多く，毎年工場への立ち入り監査も行っている。日本国内の工場への立ち入り監査も行われている。

・CSRコミュニケーションをビジネスにしている

　トッパンの特徴の1つといえるのだが，トッパンは顧客企業のCSRレポートやCSRサイトの編集・制作をビジネスにしている。年間数十社から多いときは百社を超える顧客企業のCSRレポートを制作する。環境関連イベントの企画運営やコーズリレイテッドマーケティングの企画も行う。最近では企業HPやYouTubeなどに載せる動画コンテンツも手掛けている。

　これらのことから，トッパンは数千社の顧客企業のサプライチェーンに組み込まれていて，顧客企業のCSRコミュニケーションにも携わっているという，CSRからみてもユニークな立ち位置にある企業といえる。

6. トッパンの選択と実行

6-1. 経営トップからの指示

　CSR元年と呼ばれた2003年に，トッパンはCSRコミュニケーションをテーマにした大きなセミナーイベントを企画した。当時はCSRに関連する書籍がほとんど発行されておらず，多くの顧客企業がCSRの情報を求めていた。そこで，2003年度内の開催を目指して翌2004年1月に1週間で合計11本のCSRセミナーを無料で公開した。

　当時の経営トップにCSRセミナーイベントの計画を報告しに行ったところ，トップからは，「セミナーイベントはぜひ進めなさい。そして，このようなイベントを開催すれば，取引先からは凸版印刷のCSRはどうなっているのかと

いう質問が必ず出るだろうから，それには必ず答えなければならない。サプライヤーであり，CSRコミュニケーションをビジネスにするトッパンは高いレベルでCSRを進める必要がある。セミナーイベント開催までにトッパン本体のCSRの方向性をしっかり組み立てるように」という強い指示が降りた。そこで，セミナーイベント開催までにCSRの方向性の整理を早急に行うこと，東京本社（千代田区神田）に勤務する管理職全員および東京と関西エリアの営業担当者に向けたCSR勉強会が行われることになった。

6-2. 誰からの評判を獲得するべきか

　広告宣伝をほとんど行わないトッパンがさまざまなステークホルダーに向けて全方位にCSR活動のアピールを行ったとしても，優良BtoC企業を超えることなどありえない。自社HPでアピールをしたとしても，アクセス数では1ケタどころか2ケタの差がつくこともある。CSRによるレピュテーションの向上やブランドイメージの向上を目指して，BtoC企業の後を追いかけたとしてもその差が縮まることはまずあり得ないのである。

　そこで，トッパンが取った方法は次の通りである。BtoC企業の後追いはせず，CSRの専門家や投資の調査機関から高い評価を得られるコミュニケーションに絞り込む。高い評価を得ることでいろいろなランキングやINDEXに選ばれることを目指す。そうすれば，結果的に注目を集めてレピュテーションが向上し，ブランドの向上へとつながるはずである。また，専門家から第三者の視点による細かな評価が行われるわけだから，トッパンの課題が何であるのかのフィードバックも得られるだろう。それがCSR活動のレベルアップにも役立つだろうと。

　以上のような取り組みによって得られた高い評価を積極的に発信することで，他のステークホルダーからの信頼も二次的に獲得して行けると考えたのである。

6-3. CSRコミュニケーションの戦術

　専門家から評価されるCSRコミュニケーションを進めるにあたって行ったことが3つある。

・専門家が納得するストーリーづくり

　理想的なストーリーはその企業の強みを活かしたユニークなストーリーだと思っている。実際，誰もが知っているような優良グローバル企業は，オリジナルの CSR ビジョンや指標をつくりそれを実践している。このような企業にインタビューをすると常に意識しているライバル企業はいないという回答が返ってくる。羨ましい限りである。

　BtoB 企業であり，かつ，海外には同様の総合印刷会社が存在しないトッパンが，優良グローバル企業と同じように企業ユニークな CSR ビジョンや指標を訴求したとしても，それが相手に伝わるかどうかは分からないのである。そこで，トッパンが行うストーリーづくりというのは，国際社会で合意された宣言やガイドラインや規格を積極的に取り入れて，トッパンの CSR 活動をそれらと上手く整合させながら進めるというスタイルである。

　分かりやすい例として，トッパンが行っている社会貢献活動の 1 つを紹介する。クラシック音楽専用のコンサートホールを所有しているトッパンは毎年チャリティコンサートを開催している。寄付先の選定にあたっては，国連ミレニアム開発目標を参考にしている。掲げられた目標の中から「教育」を抽出してそれに合致する寄付先を選んでいる。その理由は，トッパンが印刷技術を核にした企業であること，グループ会社には幼児教育や小中高校用の教科書出版をしている企業もあるためである。具体的な対象者の選定にあたっては識字率の低い地域で長年に渡る内戦などから難民生活を余儀なくされている子どもや母親たちとした。トッパン社員が直接訪問できる地域ではないために，現地で支援活動を行っている国連機関を通じて実施されている。

　これらのストーリーづくりは，専門家に理解されやすいことに加えて，トッパンの従業員が周囲に説明する時にも有効である。情報開示ツールの代表格である CSR レポートでは，グローバルコンパクト，国連ミレニアム開発目標，ISO 26000 等などの国際社会で合意されたものを積極的に採用しながら活動報告を行っている。

・網羅的情報開示も重視

　コミュニケーションのトレンドは常に変化している。2015 年のトレンドは，

GRIガイドラインG4や統合報告フレームワークが示すようにマテリアリティを特定して重要な情報に絞り，それら特定した重要な情報については詳細に報告を行う方向へと向かっている。マテリアリティはこれまでも言われて来たが，特にそれが主流になってきている。

ところがその一方で，ダウジョーンズやFTSE4Goodなどのサステナビリティのインデックスやブルームバーグを初めとする金融情報サービス会社は100項目以上に及ぶ評価指標を設定して調査を行い，その結果を投資家へ提供している。指標の数は年々増えて質問内容のレベルも上がっている。

つまり，国際社会で議論が行われる表舞台ではマテリアリティへ向かいながら，舞台裏では詳細情報の網羅的チェックも続けられているのである。トッパンは専門家向けのコミュニケーションに絞った段階で網羅的情報開示にも対応できるような情報開示を継続しているのである。

・第三者保証やマテリアリティ抽出のプロセス

最後が第三者意見や保証に向けた積極的な取り組みである。先述の専門家が納得するストーリーづくりや網羅的情報開示にもつながるのだが，専門家がどのような情報を求めていて，かつ彼らの求める基準をクリアするためにはどのレベルの活動が必要なのだろうか。いくらたくさんの情報を開示できたとしても，彼らの要求レベルを満たしていなければ意味がないのである。

そこで取り組んだのが，社外有識者とのダイアログ，第三者意見執筆者との時間をかけたミーティング，監査法人による保証書に向けたデータチェック，さらには，CSR調達の監査などを可能な限り積極的に受け入れるということだ。多様な視点からのチェックをテコにして，次にどのようなCSR活動を計画実行して行くべきか，そして，これらの情報開示をどのようにしていけばよいのかを判断しながら進めている。

トッパンがさまざまなランキングでトップクラスを獲得した時の手段が，多様な視点からのチェックを受けてそれらに応えるという方法だったのである。併せて重要であったのが社内各部門からの協力を取りつけることであった。今ではもうあたり前のことになっているが，今から10年前の2005～2006年当時，人事情報を社外に公表するという発想そのものがなかったのである。入社3年

後の離職率や解雇の理由，管理職の男女比率，出産後の職場への復帰率などを公開するように働きかけて，人事部や総務部からの理解と協力を得られたことが大きかった。

　トッパンが国内外から高評価を獲得できた頃というのは他社と比較してより多くの情報を開示できたことが大きい。当時の審査結果の総評を読むとCSR活動に関する評価以上に，人事情報を中心に充実した情報を開示したことに対する好意的なコメントが多かったのである。

7. CSRコミュニケーションの変化

　最後に，BtoC企業も含めたこれからのCSRコミュニケーションについて触れておきたい。

　CSR元年と呼ばれた2003年当時から，CSRコミュニケーションはパフォーマンスと同様に重要であるといわれ，特にレピュテーションを左右するといわれてきた。そしてこれからもCSRコミュニケーションが重要であることは変わらないが，時代と共にコミュニケーションの中身は変化している。従来からの「評判」というイメージによる評価から，より具体的な判断ができる材料へと移ってきている。投資先や取引先としてみた場合のより具体的で正確な判断ができる材料を求めているのである。

・特に伝える相手としての「従業員含む一般生活者」，「投資家」，「調査機関」

　これまではCSR活動が全てのステークホルダーに向けて展開されることから，CSRコミュニケーションの対象もマルチステークホルダーと考えられていた。国内企業では2000年代後半まではその傾向が強かったと思われる。実際に多くの企業のCSRレポート作成に携わるトッパンでは，毎年いろいろな企業からのオリエンテーションを受ける。その時に手渡される資料には，対象はマルチステークホルダーと書かれていることが多かった。

　ところが，特に情報を伝える必要があるステークホルダーは誰なのかについて変化が起きている。主な要因は2つあると考えている。1つ目は，GRIガイドラインG4や統合報告フレームワークによる影響である。2つ目は，

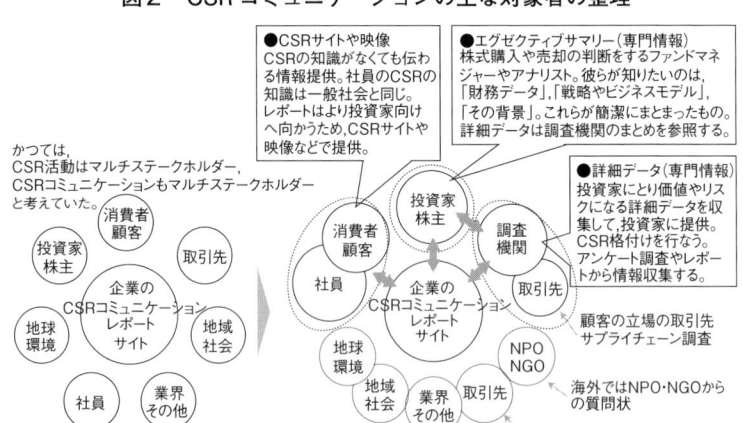

図2 CSRコミュニケーションの主な対象者の整理

Bloombergをはじめとする金融情報サービス会社の影響力が増大しているためである。

　図2はそれを表したものであるが，これからの企業は特に伝えたい相手として，「従業員含む一般生活者」，「投資家」，「調査機関」の3グループで整理することになるだろう。

　「従業員含む一般生活者」とは，国内企業が以前から伝えてきたメインの対象者にあたる。将来の顧客になりうる一般生活者，そして，まずは従業員へ啓発したいという想いは依然根強い。彼らはCSRやESGの専門家ではない。そこで，CSRの知識が少ない相手にも伝わるコンテンツを提供する必要がある。海外企業と比べて日本企業のCSRレポートやCSRサイトに特集が多いのも実はこのあたりが理由になっている。他にも，スマートフォンサイト，動画，SNSなどのICTの向上に合わせたコミュニケーションが増えていくと思われる。

　「投資家」とは，主に中長期に株を保有する機関投資家を指しているが，統合報告のフレームワークはまさにここを目指している。投資家側が最も知りたいのは環境や社会関連の細かなデータではなく，財務の主要データ，経営戦略やビジネスモデル，そして，それらを特定した背景＝理由である。GRIガイド

ラインG4が主張するマテリアリティの特定プロセスというのも，戦略やビジネスモデルが決められたその背景を説明するためのものと言い換えると分かりやすいだろう。

それでは，なぜ投資家には環境や社会関連の詳細データが必要ないのだろうか。リスク回避の観点からも実はとても重要な情報といえるのだが，この部分に関しては「調査機関」による評価結果を参考にできるということなのだ。ダウジョーンズやFTSE4Goodなどのインデックス用のアンケート調査は以前から有名でこれらの調査を実施する機関の存在も有名である。そして，今後はさらにBloombergに代表されるような金融情報サービス会社が企業のCSRに関する情報（投資家向けにはESGだが）の提供サービスを充実させてくるということだ。株式上場，未上場に関係なく，国内でCSRレポートを発行しているすべての企業のCSR情報が100項目以上のESG指標によってチェックされて既にデータベース化されている。Bloombergなどと契約しているアナリストやファンドマネージャーならいつでも自分の所有している端末からESG情報を引き出せるのである。

かつての日本企業は国内のSRI市場がほとんど成長していないことから，一般向けのCSRコミュニケーションを展開しながら，片方でダウジョーンズなどのアンケート調査に回答することを行ってきた。しかしこれからは，図2で示したように3つの方向に向けてCSRコミュニケーションを立体的に組み立てて展開することが必要になってくるだろう。

8. まとめ

凸版印刷を事例として，BtoB企業のCSRとブランド価値向上をCSRコミュニケーションの切り口から紹介してきた。そして，最後の節ではBtoBに限らずBtoC企業も含めてこれからのCSRコミュニケーション展開の姿を紹介した。

最後にトッパンが進めてきたCSRとブランド価値向上の施策をその手順に沿ってまとめておく。

・トッパンがCSRに取り組む理由をよく吟味し，CSRが必要と判断した。

- BtoB企業の特徴，印刷業界の特徴を把握しトッパンの立ち位置を確認した。
- CSR活動を推進しながらCSRによるブランド価値向上を併せて検討した。
- 誰に評価されることがブランド価値を高めるのになるのかを絞り込んだ。
- コミュニケーションの方法を決めたらブレずに継続実行していった。

その結果，広く一般社会には知られなくても，CSRの専門家や投資の調査機関から高い評価を獲得し，ビジネスパートナーからも信頼を得たのである。そして，その他のステークホルダーにもこの評価結果を2次利用して評判を上げる材料にしたのである。

<参考文献>
凸版印刷（2004, 2005, 2006, 2007, 2008, 2009, 2010, 2011, 2012, 2013, 2014）CSR報告書分析レポート
――（2004）CSRコミュニケーション―信頼とクオリティのネットワークづくり
――（2005）CSRコミュニケーション―コーポレート・コミュニケーションの展望
――（2014）CSRレポート
　http://www.toppan.co.jp/library/japanese/csr/files/pdf/2014/csr 2014.pdf
――CSR活動サイト http://www.toppan.co.jp/csr.html

4 キリンのCSVの取り組みについて
―― 「ブランドを基軸とした経営」による社会と企業の持続的な発展に向けて

太田　健
キリン株式会社 CSV 本部 CSV 推進部主幹

四居美穂子
キリン株式会社 CSV 本部 CSV 推進部

【要旨】
今企業が社会から求められていることとは何だろうか。本稿では，キリングループが掲げている「ブランドを基軸とした経営」が目指している方向性を明らかにし，企業がお客様や社会と共有できる価値創造に向けた取り組みについて述べたいと思う。キリンはマイケル・E・ポーターが提唱しているCSVの3つのアプローチを活用し，社会的価値と経済的価値の両立を目指している。

キーワード：キリン，CSV，共有価値，ブランド，マイケル・ポーター，フィリップ・コトラー，イノベーション，飲料メーカー

1. キリン株式会社設立を契機に「ブランドを基軸とした経営」へ

1-1. キリン株式会社設立の経緯

2021年のキリングループのありたい姿を描いた長期経営構想「KV2021」を実現するために，キリンホールディングスは2013年1月，国内にある3つの飲料事業会社，キリンビール・キリンビバレッジ・メルシャンを束ねた日本綜合飲料事業会社，キリン株式会社を設立した。社会の潮流に目を向けてみると，少子高齢化による人口の減少や価格競争の激化など，国内の酒類・飲料市場は今後も厳しい環境が続くと予想される。さらに，企業のあり方として従来から

の顧客満足度を向上させるだけの取り組みでは十分ではなくなり，今やどのようなミッションやビジョンも持っているかで企業が選ばれる時代に突入している（コトラー，2010）。日本のような成熟市場において，競合と戦うためには個々の事業会社が持つバラバラな商品群では企業全体のブランドを効果的に育成することができず，企業好意度の向上につなげることが難しい。このような背景により，新会社の設立に至った。

1-2.「ブランドを基軸とした経営」とは

　新会社では，お客様への「ブランドの約束」として掲げた『「飲みもの」を進化させることで，「みんなの日常」をあたらしくしていく』を行動の拠りどころとし，「ブランドを基軸とした経営」を掲げている。2015年からはさらにビジョンを具体化させ，「Quality with Surprise」というキーワードに基づき，高いクォリティでお客様にうれしい驚きや感動などの「サプライズ」のある価値を提供するよう全社で取り組んでいる。では，これらの理念を基盤として，どのような枠組みで企業ブランド形成を推進していくのか。まず，ベースとなるのは企業文化を基盤とした価値創造に向けた「組織能力の強化」である。これを形成するためにはキリンが目指すビジョンをいかに社内に向けて浸透させられるかが鍵となる。従業員が同じベクトルに向かい，お客様に新たな価値を提供する「商品ブランドの創出」と，事業活動を通じて社会的な価値を創造する「CSVの実践」の両輪により，事業を持続的な成長軌道に乗せようという戦略である（図1）。

　この両輪の機能を強化するべく，間接部門では2つの本部が新設された。1つ目は「R&D本部」である。今まで事業会社ごとに分散していた研究部門を一本化させることにより，独自技術や知識を集約し活かすことによって，楽しく新しい商品をスピーディーにお客様にお届けすることができる。2つ目が「CSV本部」である。CSV本部の最大のミッションは，日本綜合飲料事業グループに企業ブランドとCSVの戦略フレームワークを導入することで，社会と共有できる価値を創造することにある。つまり，お客様のニーズや社会課題に向き合いながら新しい商品やサービスなどを創出し，それらをお客様に発信

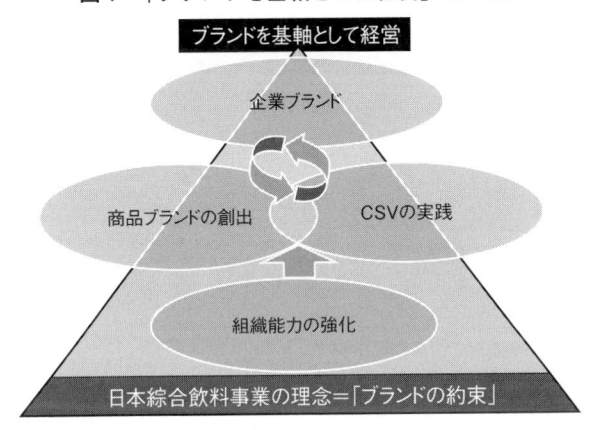

図1 「ブランドを基軸とした経営」と CSV

[図：ピラミッド図。頂点「ブランドを基軸として経営」、上層「企業ブランド」、中層「商品ブランドの創出」「CSVの実践」、下層「組織能力の強化」、基底「日本綜合飲料事業の理念=「ブランドの約束」」]

することで共感をいただき，企業の持続可能性につなげていこうとする考えである。

2. キリンの考える CSV とは

「ブランドを基軸とした経営」を推進するために CSV（Creating Shared Value）は必要不可欠な経営コンセプトであるが，本概念はハーバード経営大学院のマイケル・E・ポーターらが提唱したものである。2011年1月に発表されたハーバード・ビジネス・レビューにおいて，ポーター教授らは CSV を次のように定義している。「CSV（共有価値）とは企業の競争力を高める政策や事業活動をすると同時に，事業を展開しているコミュニティの経済的・社会的状況を発展させるものである」（ポーター，2011）。前述のように，キリンの CSV も社会的価値と経済的価値の両立により社会と共有できる価値創造を目指していくという点においては，ポーター教授らの概念と同様の捉え方をしている。

ただし，近年，CSV に関心を寄せる企業が多いことから，CSR と CSV の境界線の曖昧さがしばし議論の対象となっている。ポーター教授の CSV も，

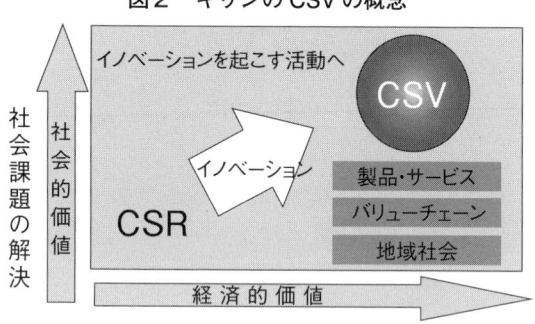

図2　キリンのCSVの概念

CSRが企業の事業活動とは無関係の社会貢献活動の総称という前提で解釈されている点が批判の的となることがある（Crane et al., 2014）。その点，2012年までのキリンでは「お客様や社会に対して価値を創造するという，企業の本来の目的追求することをCSRと捉える」としていたため，既にCSVに近い概念だったといえる。しかし現実として，企業活動の中で，社会的価値と経済的価値をKPI（Key Performance Indicators：重要成果指標）として計測し，区分することは非常に難しい。キリンでは「これはCSVか，それともCSRか」を判別するのではなく，CSV推進部がマーケティングや営業をはじめとする各部門と協働しながら，従来の社会貢献活動により経済的価値が生じる仕組みを付加したり，新たな商品やサービスには社会からの共感がより得られるような工夫をしながら企業の持続可能性を高める活動をしている。

3. CSVの3つのアプローチによるキリンの事例

キリンが取り組むCSVの重点テーマは，人権・労働，公正な事業敢行，環境，食の安全・安心，人や社会のつながりの強化，健康の増進の6つである。特に後者2つは「飲みもの」の会社であるキリンが世の中に貢献できることとして，「キリンならではのテーマ」として設定している。これらをCSVの3つのアプローチである，製品・サービス，バリューチェーン，地域社会（ポー

ター教授の解釈だとクラスター）を通しキリンらしいイノベーションを付加することで，社会との共有価値の創造を目指している（図3）。

次に3つのアプローチによる事例を紹介していこう。

3-1. 製品・サービスのキリンの CSV 事例
3-1-1. 福島の農業の応援とお客様と福島を結ぶ活動

「キリン 氷結 和梨」（期間限定商品）は，2014年11月に発売され CSV を体現する製品となった。開発の発端はキリンビールマーケティング株式会社の植木前社長が，東北復興支援の被災地の産品の販売会で佐藤前福島県知事が懸命に福島県産の農産品の PR をされている姿を見て，「何かキリンでもお手伝いができないか」と考え社内に持ち掛けたことである。その年に採れた福島県産の和梨の果汁を使った氷結は，福島の農業を応援するとともに，福島の農産物の豊かな恵みとおいしさを伝えたい，との想いでつくられた。福島県と県民の皆様が一体となり「新生ふくしま」の創造に向けた気運醸成のためのロゴマーク「ふくしまから はじめよう。」をパッケージに掲載。福島を応援したいと考えているお客様がお求めになることで復興の応援ができると大変好評をいただき，キリンが社会的な課題に真剣に取り組んでいることに対して，お客様から

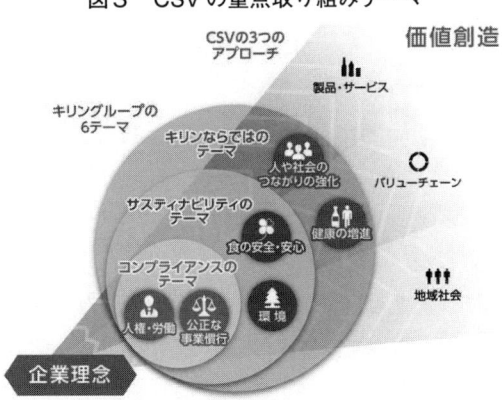

図3 CSV の重点取り組みテーマ

200　II　企業ケース /Cases of "CSR and Strategy"

図4　ロゴマーク「ふくしまからはじめよう。」をパッケージングに掲載した商品

図5　おいしさと機能を兼ね備えた商品

共感をいただいた（図4）。

　お客様からの後押しをいただいて，2015年3月には「キリン 氷結 福島産桃」（期間限定商品）を発売し，これも好評のうちに販売を終了した。

3-1-2. 独自技術を活用した製品の発売

　独自技術を活用した，おいしさや楽しさといった飲みもの本来のベネフィットに健康を加えた「KIRIN Plus-i」の製品も健康的な生活をおくりたいという社会的な課題の解決に役立っている。免疫細胞の指令塔を活性化するプラズマ乳酸菌を配合した，「キリンまもるチカラのみず」や，トクホ[1]では，「キリンメッツコーラ」がヒットしている。「キリンメッツコーラ」はトクホ史上初の脂肪の吸収を抑えるコーラ系飲料，カロリーもゼロであるため，コーラと脂肪分が多い食事の食べ合わせを諦めていた大人達を中心に支持されている。アルコールの分野でも，2015年1月に発売された「キリン のどごし オールライト」は糖質ゼロ・プリン体ゼロ・カロリーオフの機能[2]を持ちあわせた製品で「体に低負担なものを気兼ねなく，手軽に飲みたい」というお客様のニーズに応える商品を開発した（図5）。

　キリンはこれまで機能系商品やノンアルコールビールを業界に先駆けてお客様に提供してきた。糖質やプリン体オフに着目した商品の展開など，技術を進化させながら，おいしさと機能を兼ね備えた商品を次々に提案し，お客様の健

康的なライフスタイルをサポートしている。

3-2. バリューチェーンのキリンのCSV最新事例
―原材料の削減による環境負荷の低減

　環境負荷の低減とコスト削減による企業の競争力の向上は，CSVと相性が良い取り組みである。キリンは独自のパッケージング技術研究所を持つ強みを生かし，以前より使いやすい容器包装の開発や軽量化に取り組んできた。キリンビールは独自形状のリターナブル壜を使っている。壜にセラミックコーティングを施す技術により，自社比率で21％軽く国内最軽量となる大壜のリターナブル壜の開発に成功し，1993年より市場に投入している。また，同様の技術により中壜も軽量化に成功，テストを経て2015年秋には全国展開の予定である。この中壜により，製造工程と物流行程のCO_2排出量を年間約930t削減[3]できるなど，環境負荷が低減されるとともに，使いやすさの向上や「エコ」な容器包装を求めるお客様と価値の共有を図ることができる。また，アルミ缶の蓋の口径を小さくすることによる缶の軽量化や，製缶工程での環境負荷が低いエコロジー（ラミネート）缶を採用している。ペットボトルでは「キリン 午後の紅茶 おいしい無糖」で使用済みペット素材100％のR100PETボトルを採用，同じくデイリーワインではワインを気楽に楽しんでいただけるよう，軽く割れにくく輸送の環境負荷も削減できる「ワインのためのペットボトル」を導入している。中壜の取り組みもこのような従来からの流れに沿って創出されたものである。飲む人に使いやすく，軽い容器包装を目指しながら，環境への影響を最小限に抑えた持続可能な容器包装の開発・採用を進めている。

3-3. 地域社会のキリンのCSV事例―地元との連携による地域の活性化

　「人や社会のつながりの強化」はキリンのCSVの6つの重点取り組みテーマの中でも，日本全国すべての県に拠点を展開しているキリンの強みを活かせる「キリンならではのテーマ」と考えられる。この取り組みが最もマッチするのが，地域を活性する活動の中でどうキリンがお役立ちできるかである。
　キリンディスティラリー富士御殿場蒸溜所は，ウィスキーを製造するのに不

可欠な水の保全と地域の活性化が結果として商品の価値を高めている事例である。蒸溜所では富士山水源涵養林のうち約43万m^{2}の森林を借り受け，森林保全活動を毎年実施するとともに，工場の敷地面積16万9千m^{2}の約半分は自然林を残し「キリン自然の森」としてお客様に開放し，富士山麓の自然に親しんでいただいている。また，市・観光協会・周辺企業で「チームごてんば」を結成し，御殿場にお客様を誘致し地域活性化に積極的に取り組んでいる。工場では富士山が世界遺産に登録されたことを機に，工場見学の映像設備を一新し，地域の観光スポットの1つとしてより多くのお客様にお立ち寄りいただけるように工夫を重ねている。

「メルシャン 椀子（マリコ）ヴィンヤード」は，遊休農地の活用による地域の活性化と高品質のブドウ栽培による日本ワインの品質向上に繋がっている。メルシャンは長野県上田市の椀子地区の遊休農地を2003年から整備し，2010年にファーストヴィンテージとなるワインを発売。以降，2011年から4年連続で「シャトー・メルシャン マリコ・ヴィンヤード ソーヴィニヨン・ブラン」が国産ワインコンクールで金賞に輝くなど品質でも高い評価を獲得している。かつては遊休農地であったこの地を地域の方とともに良質なぶどうを生産する畑として再生し，地域の活性化にも貢献している。

ビール事業を通じた地域との協働の事例としては，県と締結した包括連携協

図6　沖縄におけるビーチクリーンアートのケラマジカ

定に沿った兵庫県の農産品の応援がある。神戸工場がJA兵庫六甲と協働で毎月開催している「ひょうごマルシェ」や，地元のスーパーマーケットとのタイアップによる県の認証食品のレシピ提案も，地産の農産品と一緒に食卓に並ぶキリンビールにできる地域活性化の取り組みである。

　沖縄での活動は，環境の保全と観光による地域活性をテーマとしている。2014年3月に慶良間諸島が31番目の国立公園に認定されたことを記念して，デザイン缶を発売した。その販売の一部（1本1円）をビーチクリーン活動の時にボランティアが着用するTシャツとして座間味・渡嘉敷の両村に寄贈した。この動きに賛同していただいた企業で「チームけらま」を結成し，11月23日に両村主催で実施された4島一斉のビーチクリーン活動を応援した。また拾ったゴミを使ってビーチクリーンアートをつくり，環境保全を呼びかけた（図6）。

4. 社会貢献活動から生まれたCSV事例

　社会貢献活動として始めた「復興応援 キリン絆プロジェクト」の支援先が事業の継続性を模索した結果，CSVとなった事例もある。同プロジェクトの活動の1つである「東北復興・農業トレーニングセンタープロジェクト」を受講された東北の農業経営者が，岩手県遠野市で「遠野パドロン」というビールに合うおつまみ野菜を遠野市の特産品に育てることを目指している。「遠野パドロン」は，ししとうのような食感で，素揚げでビールのおつまみにすると非常においしい。スペインでは日本の枝豆と同じ定番の組み合わせとのことで，2014年夏にビアレストラン「キリンシティ」の全店舗で季節メニューとして提供し，大好評となった。農業と地域の活性化とキリンの両方の価値向上につながっている。

5. 最後に

　ご紹介した事例は従業員に自社の身近な取り組みを知ってもらうことでCSVを理解するためのツールとしても活用されている。前述のように「ブラ

ンドを基軸とした経営」の実現には「組織能力の強化」が基盤となるが，実際のところ従業員一人ひとりが異なる役割を担う中で，全員がCSVを理解し推進している状況を創りだしていくには時間を要する。最近の社内調査によれば，約60%の社員が「CSVを経営コンセプトに掲げていることに共感している」と答えており，社内で一定の理解は得られている。しかしそれが必ずしも「CSVを実践できる」と同義語であるとは限らない。つまり，理解と行動の間にある壁を越えなければ，お客様から共感をしていただけるような商品やサービスの継続的な創出は期待できない。キリンのCSVはまだ始まったばかりではあるが，従業員一人ひとりが何らかの形で社会課題との接点を持ち，日常業務の中で解決を意識し実践することで，お客様に長く愛される企業になることをこれからも目指していきたい。

(1) 特定保健用食品：からだの生理学的機能などに影響を与える保健機能製品を含む食品で，脂肪の吸収を抑えることに役立つ，などの特定の保健の用途に資するもの。
(2) カロリー，糖質については栄養表示基準，プリン体については自主基準による表示。
(3) 年間1千万本製造した場合。

<参考文献>

フィリップ・コトラー，ヘルマワン・カルタジャヤ，イワン・セティアワン著，恩藏直人監訳，藤井清美訳（2010）『コトラーのマーケティング3.0　ソーシャル・メディア時代の新法則』，朝日新聞出版。

マイケル・E・ポーター（2011）「戦略と競争優位」，『ダイアモンド・ハーバード・ビジネス・レビュー』，2011年6月号。

Crane, A., Palazzo, G., Spence, L. J. and Matten, D.（2014）'Contesting the Value of the Shared Value Concept', *California Management Review*, Vol. 56/2.

Ⅲ 投稿論文（査読付）／ Reviewed Articles

1 Psychological Empowerment for Competitive Advantage
　　—A Resource-based Approach to Human Resource Management in Hospital Industry of Nepal
　　　　　　　Sunita Bhandari Ghimire and Dhruba Kumar Gautam

2 高齢化地域の持続可能性に資する地域企業のイノベーション戦略
　　—徳島県上勝町「いろどり」からの考察　　　　芳賀和恵

3 韓国の社会的企業の'制度化を通じた'育成に関する考察
　　—「社会的企業育成法」を巡る現状と課題を中心に
　　　　　　　　　　　　　　　　　　　　金　仁仙

4 消費財の情報特性がCSR活動に与える影響の分析
　　　　　　　　　　　　　　　　　　　　吉田賢一

1 Psychological Empowerment for Competitive Advantage
—— A Resource-based Approach to Human Resource Management in Hospital Industry of Nepal

Sunita Bhandari Ghimire
Assistant Professor, Tribhuvan University, Kathmandu, Nepal

Dhruba Kumar Gautam
Associate professor, Tribhuvan University, Kathmandu, Nepal

[Abstract]

In this eve of competition, to ensure organizational survival, organizations should increase their competitive power. And one reliable method to increase competitive power is the effective use of available resources whether resources can be human or capital. Based on these, this study has been carried out to find out the real situation of empowerment of employees in the hospital sector of Nepal for the purpose of exploring whether they are able to increase their competitive advantage through employee empowerment or not. The study has used two well-grounded theories to measure the relationship between empowerment and competitiveness with resource based approach to human resource management. Menon's (2001) three component model of psychological empowerment and Wright et al.'s (1994) four component model of competitive advantage has been used for finding the situation of empowerment as well as relationship between empowerment and competitive advantage. Convenient and Judgmental sampling has been followed. Four hospitals from both public and private sectors have been selected purposively. Altogether 120 employees have been defined as samples and equal numbers of questionnaires have been distributed to all selected organizations. Descriptive, correlational and regression analysis has been used to achieve research objectives and to test the proposed research hypothesis. In total 24 items has been used for measuring these different variables. This study has followed the quantitative study with co-relational, descriptive and analytical research design. To describe the responses, mean value of each variable has been computed and tested with standard deviation for variance test and the Cronbach's alpha for testing reliability has been calculated with the help of SPSS 18. Correlation has been calculated to measure the relationship between different variables. Multiple Regression analysis has been conducted to examine the relationship between studied variables and to identify the strength of that relationship.

Key words: Empowerment, competitive advantage, Resource based approach, Psychological empowerment

1. Background

Organizations of this era are facing a high level of competition as both the opportunities and threats are increasing due to different issues like globalization, liberalization, outsourcing and so on. To ensure organizational survival, organizations should increase their competitive advantage and one reliable method of increasing competitive advantage is the effective use of available resources where resources can be human based or capital based. Competition is an issue of service and product. The question then is, what is the factor that has a major role in both of them? Schuler and Jackson 1987 have shown that there is a linkage between competitive advantage, HR practices and performance. Also HR practice results from different human resources existing within the organization.

According to different researchers as Cappelli and Singh, 1992; Pfeffer, 1994; Ulrich, 1991; Wright and McMahan, 1992; gaining competitive advantage through human resource has become an important focus of research and analysis in the Human resource management field (Bae and Lawer, 2000). As quoted by Bay and Lawer, 2000, to increase the competitive advantage in Korea i.e. Asian Countries restructuring has been focused which has led to the introduction of "New human resource management (NHRM). NHRM approaches involve greater reliance on teams, empowerment and performance based evaluation, pay and staffing. According to them, high-involvement

HRM strategy starts with management philosophies and core values that emphasize the significance of employees as a source of competitive advantage. According to them, Huselid's point of view of high-involvement HRM strategies may enhance, reinforce and sustain both the competence and commitment of employees, essential for competitive advantage in today's turbulent business environment. Hospital sector cannot be the exceptional case being an industry with large number of hospitals competing with each other for the long term sustainability all over the world including Nepal.

Achieving competitive success through people involves fundamentally altering how we think about the workforce and the employment relationship. Firms that take this different perspective are often able to successfully outmaneuver and outperform their rivals (Pfeffer, 2005). While there is unanimous agreement that firms sustain competitive advantage from many different and unique resources, it is evident that the majority of leading firms have one unique resource in common (Barney & Wright, 1998; Pfeffer, 1994; Wright et al., 1994). The HR practices would therefore include (Schular and Jackson 1987, cited by Boxall, 1996) selecting highly skilled individuals, giving employees more discretion, using minimal controls, making a greater investment in human resources, providing more resources for experimentation, allowing and even rewarding occasional failure, and appraising performance for its long-run implications.

Based on above discussion, it can be articulated that psychological empowerment and organizational competitive advantages are context specific, which is not exceptional in Nepalese Hospital sector. Although the country-Nepal has entered into liberalization and privatization process, very few decision makers and academics are acquainted with dynamics of new HRM concepts and practices. Nepalese decision makers are now being faced with a number of challenges brought by both internal and external contexts. Externally they have to cope with many unexpected changes undergoing in external environment due to rising competition especially in service sector, unstable political system, lack of proper institutional mechanism for enforcing rules and regulations, rising unemployment and low growth rate. Similarly, internally companies have to take number of initiatives to raise capability of employees making them ready to provide delivery to cope with external challenges. Therefore, organization need to invest in resources to develop their HR competent to face the challenges of twenty-first century (Gautam, 2015). In this paper we have taken Hospital as a field of study.

According to Gautam (2015), despite the importance of Human Resources in increasing the competitive advantage, evidence from developing countries (particularly Nepal) is lacking. Review of the past two years' research contribution in the South Asian Journal of Global Business Research (SAJGBR), a journal dedicated to advancing international business theory by focusing upon South Asia, indicates that less than 3 percent of research focused upon evidence from Nepal; quoted from Khiljii, 2013. Hence there is a need to study HR issues in Nepal. Strategic HR management in context to Nepal can be said as infant stage where only 56 percent organizations had a written strategy (Adhikari and Gautam, 2007). The researchers tried to find the research article related to this field but there is dearth of research articles which could provide the insight in the empowerment of the health workers for increasing the competitive advantage of hospitals.

Human resources can contribute a lot to make organizations more competing and raising value of the firm by making changes at the workplace (Adhikari, 2012). Empowerment in Nepal for service sector industries has not been in the priority. Majority of Nepalese business executives are unaware of the employee empowerment concept and those who are aware too have low comprehension of the concept. These all show that empowerment of employees is necessary in Nepal to increase the competitive advantage of the organizations. Therefore, this study aimed to explore the real situation of psychological empowerment in Nepalese hospitals examining the relationship between psychological empowerment and competitive advantage. It further, aimed to analyze the elements essential for increasing competitive advantage in the perception of employees of hospitals.

2. Overview of Nepalese Health Industry

It is very difficult to identify the exact date of hospital development in Nepal as many areas of the Nepalese history are missing. Nepal has its' own indigenous system of medicine which remained somehow mainstream health system till initial days of modern Nepal. The history of health and hospital development dates back to the ancient Nepal or Lichchhavi period. In the reign of the Amshu Verma (605-620 AD) one of the historical document found in 604 AD has mentioned about the Aarogyashala or hospital (Ayurvedic), but no elaborate explanations has been found.

Tenure of the Prime Minister Bir Shamsher is remembered as the landmark for the health and hospital development in Nepal. During his premiership the first hospital of the country in modern medicine was established in Kathmandu in the year 1889 AD as Bir Hospital. The regionalization plan of health services in public sector initiated in 1960s seems weakened as no comprehensive care hospitals such as zonal and regional hospital have been established. The health facilities are concentrated in few cities looks imbalance development of health sector and some districts do not have hospitals till date. Most of the hospitals are centralized in Kathmandu valley- the capital of the country. According to department of health services, there are only 8 central hospitals and 10 zonal hospitals established by government of Nepal. A report on census of private hospitals in Nepal reported that 214 private hospitals are working in Nepal. Report of ministry of health and population identified approximately 54,177 health worker, with 32,809 in the public sector and 21,368 in the private sector.

The Human Resources for Health (HRH) situation in Nepal has been met with several key challenges particularly related to the shortage and uneven distribution of the health workforce in the country. A trained and skilled health workforce at the right place with adequate motivation and support are crucial for achieving the MDG targets by 2015. The performance of the health workforce plays a crucial role in the improvement of health outcomes, due to its impact on accessibility to health services and appropriateness of care provided to service users. There is increasing global consensus for the need to consider the health system in its entirety, taking into consideration the limitations of public health budgets and the use of the private sector as a support in the struggle to provide higher quality services to a greater number of people. The low level of motivation among health workers has been identified as a key issue in the current human resources crisis in the health sector. Yet the focus on motivation and performance of health workers through improved working conditions is often overlooked by governments in favor of macroeconomic issues. In Nepal, ensuring a comprehensive strategy that maximizes health worker motivation is crucial (Shrestha and Bhandari, 2012).

3. Rationale of the Study

Resources are the most important for operating an organization efficiently and effectively. Different resources including human resources are to be utilized in such a way that they can create the competitive advantage in the organization. Resource based theory states that resources in organizations should be rare which can add value to organization, un- substitutable as well as inimitable so that they can create competitive advantage to the organization. Human resources are heterogeneous in nature due to difference in background which if utilized effectively in organization could create competitive advantage for the organization.

Resource Based View (RBV) suggests that the method in which resources are applied within a firm can create a competitive advantage (Barney, 1991; Peteraf, 1993; Wernerfert, 1984). The resource based view of the firm suggests that an organization's human capital management practices can contribute significantly to sustaining competitive advantage by creating specific knowledge, skills and culture within the firm that are difficult to imitate (Barney, 1991; Wright et, al., 1994). In this context we can work to find out: Are hospitals empowering their personnel from resource based approach of human resource management?

The following sub-questions derived from different variables under study are as follows:

1. What variables are given importance for psychological empowerment in hospitals?
2. What is the relationship between psychological empowerment and competitive advantage in hospitals?

4. Focus of the study

The nature of services as being intangible, heterogeneous, perishable, produced, and consumed at same time makes it peculiar to deliver, and challenging to organizations to achieve a differentiation from the others. Health workers therefore, become the voice and face of the hospital, but it is not enough that they be trained to provide quality service, that they know what to do and how to do it. It is also essential that they have the requisite authority to make decisions regarding Patient satisfaction. This is one of the arguments for health worker empowerment since they act as an interface between the patient and the hospital (Kabene, 2006).

Hospitals are the very crucial organizations where the skills of the employees play a vital role in the success or failure of an organization, so effective management of the available resources is essential in the Hospitals likewise the other sectors. This is the era of competition so it is essential that hospitals have to increase their competitive advantage (Operational research report, 2012). Human resources is that resource which being rare, inimitable, un-substitutable, and can add value being a rational animal has the potential to increase the competitive advantage of an organization (Barney, 1991; Wright et al., 1994).

Nepalese organization and people working in those organizations responsible for policy making are generally not convinced about the benefits of investment in human resources (Operational research report, 2012). Nepalese researchers and academicians have not considered these things as the field of research especially in the field of hospital. This study is focused in the psychological empowerment and competitive advantage of the hospital sector of Nepal.

5. Limitation of the study

There are few limitations in the present research. Some of the major limitations are listed below:
(1) Only limited hospitals among several hospitals within the industry are taken into consideration.
(2) Different variables are taken by different scholars at different times for the study of empowerment and competitive advantage. Some variables that do not seem to be applicable for this study are not taken into consideration.
(3) Study and respondents are limited in the organizations within Kathmandu valley. This study does not represent the outsider's views.
(4) Study is totally carried out on the basis of primary data collected through Likert Scale type from scaling 1 to 5 which may not necessarily capture the actual feature of the real situation.

6. Review of literature

Over the last two decades, 'empowerment' has attracted the interest of many organizational theorists and management practitioners (Argyris, 1998, Conger & Kanungo, 1988; Lashley, 1994; Neilson, 1986; Thomas & Velthouse, 1990) as a concept which appears to offer tremendous potential in the enhancement of organizational effectiveness.

Empowerment is the effective way of developing human resources for increasing the competitive advantage of an organization. Conger and Kanungo (1988) define empowerment as a process of enhancing feelings of self-efficacy among organizational members through the identification of conditions that foster powerlessness and through their removal by both formal organizational practices and informal techniques of providing efficacy information.

In research on healthcare quality management, strategic HRM commands paramount importance for program results and sustainable competitive advantage (Zairi, 1998). Successful implementation of employee empowerment and team-building is essential for healthcare quality programs (Adinolfi, 2003). Furthermore, strategic HRM in healthcare has proven to be imperative for a sustainable competitive advantage (Kanji and Sa, 2003). Hospital quality management and HRM practices can be expected to enhance corporate competitive advantage. Appropriately designed program practices could improve the value and rareness of programs through greater efficiency in the implementation of quality management practices (Charles, 2006). Hospitals need to employ and coordinate specialized knowledge, skills and abilities embedded in their employees to deliver quality care to patients (Wiig, 2002; Van Beveren, 2003; Yavas and Romanova, 2005).

There is increasing evidence in support of the potential contribution of HRM to hospital functioning and patient outcomes (Gowen et al., 2006; West et al., 2006). Specifically, research in Australia by Bartram et al. (2007) found some support for the relationship between strategic HRM and improved organizational outcomes in hospitals and community health centers. Similarly, West et al. (2006) found that, after controlling for confounding factors, greater use of a complementary set of HRM practices has a statistically and practically significant relationship with patient mortality. However, in spite of the potential contribution of HR and HRM to organizational and patient outcomes, the hospital HR function has been found to be underdeveloped and lacking capacity in both Ireland and the UK (Fitzgerald et al., 2006; Hyde et al., 2006; National Health Strategy Consultative Forum, 2001). Although similar challenges have faced HR across several industries (Kamoche, 1994), it has been suggested that the potential for the development of a strategic approach towards HR and HRM in healthcare is uncertain. Rationales for this include the historic skepticism associated with hospital management, the limited credibility of the HR function, its focus on a narrow operational contribution and its peripheral position in the organization, (Bach, 1994; Barnett et al., 1996).

Sparks, Bradley and Callan (1997) reported that employees who are fully empowered and communicate with customers in attentive manner could evoke more customer satisfaction. Farrell, Souchon and Durden (2001) indicated that customers' perceptions of service quality would be based almost entirely upon the service behaviors of employees. Customers specially appreciate the service encounter while measuring service quality, therefore service behaviors of employees reveal more important in the service delivery process. Here the employees are health workers and the customers the patients.

As quoted from Khan et al., according to Thamizhmanii and Hasan (2010) the work force is referred as front liners. They are like armed soldiers at the war front. Zhengliang and Yilei squeezing from studies (of Gronroos, 1981; and Chebat & Kollias, 2000) proposed that front-line employees need to be empowered to make an immediate decision, otherwise they will miss sales opportunities or opportunities for improving service quality. Samad (2007) concluded from studies that researchers and leaders worldwide have called for the empowerment of employees to help their organization to compete successfully in the highly competitive marketplace (Khan et al., 2011).

When talking about empowerment in Health care sector, Human resources can be defined as the different kinds of clinical and non-clinical staff responsible for public and individual health intervention (Kabene, 2006; Alemu, 2011). Health care workers are required to ensure that the workforce is aware of and prepared to meet a particular country's present and future needs. A properly trained and competent workforce is essential to any successful health care system (Kabene, 2006). Many health workers in developing countries are underpaid, poorly motivated and very dissatisfied according to Kabene which shows lack of empowerment.

According to Kabene et al., different systems are followed in America, Germany and Canada for empowering Health care personnel so that the organizational competitive advantage could be increased. However in developing countries, the problem of investing in the training of health care professionals, thus using precious national resources but also losing many of their trained professionals to other areas of the world that are able to provide them with more opportunities and benefits (Kabene, 2006) are losing their competitive advantage.

In a dynamic environment, health managers need to combine leadership, entrepreneurial and administrative skills to meet the challenges that the changing socio-political, economic and technological landscape presents as well as the expectations of patients, health professionals, politicians and the public. Without good management we will be unable to improve efficiency, effectiveness and responsiveness in the delivery of health services or upscale interventions to achieve health goals (Alemu, 2011).

Strategic skills, task related skills, people related skills, self-management skills are essential for the employees related to health care. Determining and assessing competencies is a vital precursor for improving professional development and the alignment of individual development with the need of an organization or profession. The competency based approach to professional development is well accepted in human resource management literature where it is regarded as a critical part of the overall management development process as it allows one to identify the gaps between current skills and the skills required (Alemu, 2011; Kabene, 2006; Dadashinasab, 2012).

According to Operational research report 2012, in the underdeveloped country like Nepal, research has highlighted the importance of three broad areas as the building blocks of HRM, as needed to maximize workforce productivity and performance which are- employees with the necessary ability, adequate motivation for them to apply their abilities and opportunity for them to apply discretionary behavior which are the essential factors of psychological empowerment.

Ministry of Health and population accords high priority for the development of competent human resources for health through various training activities (Operational research report, 2012) so that health sector can develop its competitive advantage.

The great interest in empowerment in the hospital industry is by large associated with the belief in its potential to enhance patient satisfaction. Besides it is widely believed that empowerment is associated with gaining and sustaining of competitive advantage in the service industry which is hospital industry in this context. (Conger and Kanungo, 1988; Bowen and Lawler, 1992; Lashley, 1995; Quinn and Spreitzer, 1997).

Randolph (1995) defines employee empowerment as "a transfer of power" from the employer to the employees. But researchers also for instance argued that empowerment is not only having the freedom to act, but also having higher degree of responsibility and accountability. This indicates that management must empower their employees so that they can be motivated, committed, satisfied and assist the organization in achieving its objectives which also implies to health workers in hospitals.

When talking about empowerment, as stated above it is always not helpful in increasing the competitive advantage of the organization. If not managed properly with proper planning it would be a factor of increasing the unwanted cost in the organization creating the burden (Spreitzer, 1996). If the employees are unable to accept the condition, the service delivery to the customers will be slow (Bowen and Lawler, 1992) creating the situation of unwanted time consuming for unnecessary work.

The greatest challenge comes when it is the time of implementation. So, it is said that empowerment is a good idea but unworkable for large corporations. Instead many researchers argue that empowerment should only be tried in small companies where the risks of failures are less. But, in this study, the researcher is trying to know the degree of relation between Psychological empowerment and competitive advantage.

Given the practical importance of psychological empowerment, and the fact that businesses in China have started to rely more on employees to enhance competitive advantages for long-term survival, researchers have started to investigate the contributing factors of enhanced employee psychological empowerment in the Chinese context (Li, 2011).

Research suggests that psychological empowerment, defined as intrinsic task motivation manifested in an individual's sense of meaning, competence, self-determination, and impact (Conger and Kanungo, 1988; Spreitzer, 1995; Thomas and Velthouse, 1990), is associated with various positive employee work outcomes. These positive relationships have been well documented in Chinese samples.

6-1. Psychological theory of empowerment

Psychological empowerment was described as based in four cognitions that affected an employees' intrinsic motivation, namely meaning, competence, self-determination and impact (Thomas & Velthouse, 1990). Spreitzer (1995) built upon Thomas and Velthouse's (1990) model and validated a measure of psychological empowerment. Meaning represents a fit between a work goal or purpose and a person's own ideas, values and beliefs. Competence can be defined as a person's belief in his own capacity to perform activities with skill. Self-determination reflects autonomy over the initiation and continuation of work behavior and processes; making decisions about work methods, pace, and effort are examples. Impact refers to the degree to which a person can influence strategic, administrative, or operating outcomes at work (Spreitzer, 1995).

Menon (2001), on the basis of the review of literature and mainly the work of Conger and Kanungo as well as Thomas and Velthouse identified three main dimensions of the experience of power as the root of psychological empowerment underlying the empowerment process as:
a) Power as perceived control, b) power as perceived competence and c) power as being energized toward achieving valued goals.

6-2. The resource-based view model

Firm resources include all assets, capabilities, organizational processes, firm attributes, information, knowledge, etc. controlled by a firm that enable the firm to conceive of and implement strategies that improve its efficiency (doing things right) and effectiveness (doing the right things). Firm resources can be conveniently classified into three categories: physical capital resources, human capital resources and organizational capital resources.

Physical capital resources include the physical technology used in a firm, a firm's plant and equipment, its geographic location, and access to raw materials. Human capital resources include the training, experience, judgment, intelligence, relationships and insight of individual managers and workers in a firm. The organizational capital resources include a firm's formal reporting structure, its formal and informal planning, controlling, and coordinating systems, as well as relations among groups within a firm and between a firm and those in its environment (Barney, 1991).

Barney 1991 suggests that in order to understand sources of sustained competitive advantage, it is necessary to build a theoretical model that begins with the assumption that firm resources may be heterogeneous and immobile. To have this potential, a firm resource must have four attributes.

・It must be valuable, in the sense that it exploits opportunities and/or minimizes threats in a firm's environment.

・It must be rare among a firm's current and potential competition.

・It must be imperfectly imitable.

・There cannot be strategically equivalent substitutes for this resource that are valuable but neither rare nor imperfectly imitable.

Firm resources can be imperfectly imitable for one or a combination of three reasons:

1. The ability of a firm to obtain a resource is dependent upon unique historical conditions.
2. The link between the resources possessed by a firm and firm's sustained competitive advantage is causally ambiguous.
3. The resource generating a firm advantage is socially complex.

Wright, et al. 1994 has shown that the human resources can be a source of competitive advantage because they meet the criteria for being a source of sustainable competitive advantage. Human resources add value to the firm, are rare, cannot be imitated and are not sustainable.

6-3. Achieving competitive advantage through empowering health workers

Kahreh et al. (2011) based on the prior literature take three dimensions-autonomy through boundaries, information sharing, and team accountability as the organizational practices associated with the empowerment climate of project teams having varying degree of vitality. These dimensions that organizations focus on and show great interest in, while providing services and products so as to meet market demand, can help organizations achieve competitive advantage.

Empowerment is perceived as a solution to highly regulated workplaces where creativity was stifled and workers were alienated, showing discontent both individually and collectively. An empowered and committed workforce is widely claimed to be essential for the effective functioning of modern organizations (Rawat, 2011) Organizations need to manage and improve the performance of their employees.

In this way employees are able to use empowerment responsibly so that in most cases they will not have to involve their managers when dealing with everyday problems. An empowered workforce will lead to achieving competitive advantage (Conger, 1988; Forrester, 2000; Spreitzer, 1997; Sundbo, 1999). Some of the benefits of psychological empowerment programs for the individual employee include confidence about their ability to perform their work well (Spreitzer, 1997) perceived control in terms of a sense of competence and self-determination (Menon, 2001) a clear understanding of their role in an organization, lower absenteeism and turnover, a sense of ownership, taking responsibility, higher levels of motivation, commitment, performance and job satisfaction. Human capital is a factor that can promote competitiveness as it provides the required skills, knowledge, attitudes and capacities for developing competitive strategies of the hospitals.

The cooperation and coordination between management and employee is crucial for effective and efficient functioning of an organization (Baniya, 2004). The interest of managing human resource is based around the notion that people at work are the key source of sustained competitive advantages. HR practices integrated internally and externally support to achieve its bottom line objective-performance. This clearly exhibits that human resources can contribute a lot to make organizations more competing and raising value of the firm by making changes at the workplace (Adhikari and Gautam, 2011) which is taken as a basis for its implementation in hospitals under this article.

7. Conceptual Framework

The study uses well-grounded theories to measure the relationship between empowerment and competitiveness with resource based approach to human resource management. Menon's (2001) three component model of psychological empowerment and Wright et al.'s (1994) four component model of competitive advantage are used for finding the situation of empowerment as well as relationship between empowerment and competitive advantage. Fig. 1 depicts the conceptual framework used to ground the study.

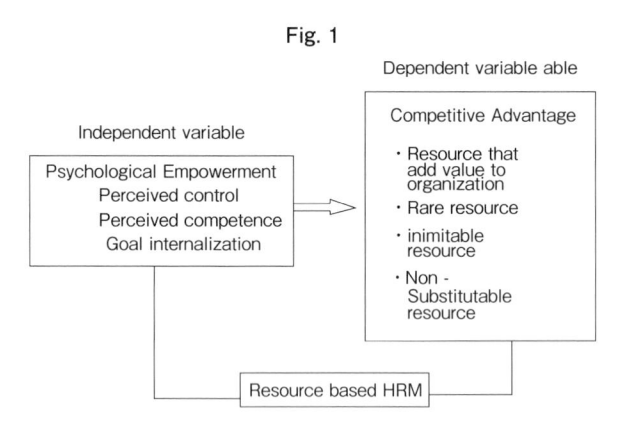

Fig. 1

```
                                    Dependent variable able
                              ┌─────────────────────────────┐
       Independent variable   │    Competitive Advantage    │
  ┌─────────────────────────┐ │ · Resource that             │
  │ Psychological Empowerment│ │   add value to              │
  │   Perceived control     │⇒│   organization              │
  │   Perceived competence  │ │ · Rare resource             │
  │   Goal internalization  │ │ · inimitable                │
  └─────────────────────────┘ │   resource                  │
              │                │ · Non -                    │
              │                │   Substitutable            │
              │                │   resource                  │
              │                └─────────────────────────────┘
              │                          │
              └──┤ Resource based HRM ├──┘
```

7-1. Psychological Empowerment

Psychological empowerment is defined as a motivational construct which focuses on the cognition of the individual being empowered (Spreitzer, 1996; Thomas and Tymon, 1994; Spreitzer, Kizilos and Nason, 1997). Perceived control, perceived competence and goal internalization are the elements of psychological empowerment which are included in the purposed study as construct developed by Menon. Empowering strategies such as delegation, increased participation and providing information and resources can lead to a sense of perceived control (Conger and Kanungo, 1988). Perceived competence is the feeling of being able for taking initiation of any task given to him (Thomas and Velthouse, 1990). Goal internalization is the ability of employees to feel as transformational leader to transform the beliefs and attitudes of employees in line with the organization's mission and objectives (Yukl, 2006, Conger and Kanungo, 1987).

7-2. Competitive Advantage

Human resource management is taken as a discipline that deals with HR activities used to support the firm's competitive advantage. Schuler and MacMillan (1983) have suggested that organization is conceiving such strategy that allows effective competitive advantage primarily through its human resources as a sustainable competitive advantage.

In this paper, researchers have defined four variables as the components of competitive advantage based on different researches (Ulrich and Lake, 1990; Barney, 1991).

7-2-1. Add value to firm

To increase the competitive advantage of an organization, some value should be added to the organization product/ service or the processes of an organization. Individuals differ in their skills and level of skills can add value to the firm.

7-2-2. Rare resources

Human resources with superior ability are rare. Organizational way of empowering employees can make them rare.

7-2-3. In-imitable resources

Well-developed human resources are generally in-imitable by competitors. Unique social relationships cannot be duplicated. Empowerment should be supported by the culture of the organization.

7-2-4. Non-substitutable resources

Human resources of an organization are to be developed in such a way that they are not easily substituted by the competitive organizations. Cultural support, technological support, leadership support and so on may be helpful in making the resources non-substitutable.

8. Formulation of Hypotheses

This study will examine the empowerment practices of Nepalese service sector hospitals. It will see the relationship between psychological empowerment and competitive advantage.

Following hypothesis can be formulated on the basis of the literature review and the conceptual framework which is proposed.

Ha: Hospitals empowering health workers with resource based approach of HRM can have competitive advantage.

Different sub hypotheses are formulated on the basis of the review of literature, objectives of the study and the research problem which will be tested on the analysis part. Hypotheses that are developed are as follows:

Hypothesis 1: There is impact of psychological empowerment in competitive advantage of hospitals.

Hypothesis 2a: Across the public and private hospitals, there is difference in organizational mean competitive advantage.

Hypothesis 2b: Across the public and private hospitals, there is difference in organizational mean psychological empowerment.

Hypothesis 3a: Across the different ages of hospitals, there is at least one type of organizational mean market share coverage that significantly differs from others.

Hypothesis 3b: Across the different ages of hospitals, there is at least one type of hospital mean competitive position in terms of sales that significantly differs from others.

Hypothesis 4a: There is significant difference in mean for competitive advantage between married and un-married health worker.

Hypothesis 4b: There is significant difference in mean for competitive advantage between male and female health worker.

9. RESEARCH METHODOLOGY

9-1. Research Design

Empowerment of the employees can be done structurally or psychologically. Perceived control, perceived competencies and goal internalization are the factors responsible for psychological empowerment of the employees. It is hypothesized that if there is empowerment of health workers within the hospitals, hospitals are able to create the competitive advantage for themselves.

This study follows the quantitative study with co-relational, descriptive and analytical research design. In this study, competitive advantage is considered as the dependent variable and empowerment as the independent variables. Demographic factors (age, gender) will be considered as the control variable.

Here, a five point Likert scale questionnaire (with 5= strongly agree to 1 strongly disagree) has been used for each of the statement. Some statements are used by the researchers to find out the factors that have been given emphasis by employees as the factors responsible for empowering them.

9-2. Sources of Data

Primary data are used for finding the relationship between empowerment of health workers and competitive advantage of the hospital. This model is based on the resource based approach of the human resource management.

Necessary data were collected by distributing questionnaire personally visiting the sample organizations. Two private and two public organizations from hospital industry are taken for study i.e. altogether four organizations are taken into consideration. These organizations were selected on the basis of convenient sampling method.

A fully structured questionnaire was developed for measuring the different variables so that proposed research can be carried out. Three dimensions of psychological empowerment and four dimensions of competitive advantage are taken into consideration.

The main objective of this study is to examine the effect of psychological empowerment in the competitive advantage of the hospitals.

In this study, competitive advantage is the dependent variable whereas psychological empowerment is the independent variables. Relation between these variables is viewed from the resource based approach of Human resource management. These variables and questions are based on the study of Menon (2001) and Wright et al. (1994). A five points Likert Scale (with 5= strongly agree to 1= strongly disagree) has been used for each of the statement which measures the different items. Demographic factors (age, gender) as well as the age of the organization are considered as potential variables that could make the result different.

9-3. Instrumentation

The data used for this study were obtained through Menon (2001) three component model and Wright (1994) four component model. Additional demographic information items have been included to facilitate the research so that exact position could be explored. The survey consists of the following measures;

・The three component of psychological empowerment is represented by 9 items to measure psychological empowerment. Three items to measure perceived control, three items to measure perceived competence and three items to measure goal internalization.

・The four component of competitive advantage is represented by 12 items to measure competitive advantage of an organization. Three items to measure value added to organization, three items to measure the inimitable resource of an organization, three items to measure non-substitutable resource of an organization and three items to measure rare resource of an organization.

9-4. Population and Sample

Mainly three major public hospitals are providing general services to stakeholders within the Kathmandu valley where as numerous private hospitals are providing the general services in the same sector. But the patients being provided the services are greater in number of public hospital in comparison to private hospital in total. Total number of hospitals in Nepal has been discussed earlier.

In this study, hospitals of Kathmandu valley of Nepal which are established at least before five years are taken as population. Two hospitals from government sector and two from private sector are selected for study as the sample.

9-5. Sample size

Judgmental sampling method was used to select samples of hospitals for study. Four selected hospitals are Teaching Hospital, Bir Hospital, Kist hospital and Kathmandu medical college where Teaching Hospital and Bir Hospital are government hospital whereas Kist and Kathmandu medical college are private hospitals. Altogether 120 health workers have been defined as samples and equal numbers of questionnaires are distributed to all selected organizations. Doctors, nurses and administrative staff are taken as samples for the defined study.

9-6. Statistical analysis

Correlation is used to measure the relationship between different variables. Regression analysis has been conducted to examine relationship between studied variables as highlighted in the research questions, and to test the research hypotheses. This study employs multiple regression analysis to investigate the relationship between variables and to identify the strength of that relationship.

9-7. Reliability Test

The items in the factorial groups were also tested for reliability. Reliability test was undertaken to ensure that the research findings has the ability to provide consistent results in repeated incidences. To check the reliability aspect of the items and its factorial groups, internal consistency analysis using SPSS was performed. The items were grouped into its respective factorial group and coefficient alpha was calculated. As suggested by Nunally (1967) value of Cronbach's alpha above than 0.6 is sufficient value in an exploratory research.

9-8. Technique of Analysis

A five point Likert Scale (with 5= strongly agree and 1= strongly disagree) which has been used to collect the views of respondents. Factor analysis was done as various items were used for collecting views as analysis of individual items was not able to make the analysis effective. In the course of factor analysis principal component analysis was used to develop factors out of the related so that further analysis could be made more convenient and reliable with wide coverage. Using the SPSS model, correlation between the items was calculated at first and the components with Eigen values greater than 1 are taken for further analysis under rotated component matrix. Items that were not able to describe the factor were dropped.

(For detail see the process of principal component factor analysis). This process is highly useful for the extraction of the variables which has been used in this research analysis.

ANALYSIS AND INTERPRETATION OF DATA:

Table 1　Factor analysis for psychological empowerment

Items	Factor	Chronbach alpha after dropping items	KMO and Bartlett's Test	Significance
1. Skills and abilities	Perceived competence	0.759	0.746	0.000
2. Competence to work effectively				
3. Capabilities required*				
4. Influence the way work is done	Perceived control	0.737	0.697	0.000
5. Influence decision in department				
6. Authority to make decision				
7. Inspired by achievements	Goal internalization	0.686	0.604	0.000
8. Inspired by goals				
9. Enthusiastic on organizational goals				

Table 2 Factor analysis for competitive advantage

Items	Factor	Chronbach alpha after dropping items	KMO and Bartlett's Test	Significance
1. Higher quality people than competitors	Rare Resource	0.713	0.715	0.000
2. Development of unique intellectual capital				
3. Employees market value				
4. Cost effective Human Resources	In-imitable resources	0.734	0.723	0.000
5. Competent and skilled human resources*				
6. Attraction of competent and skilled persons				
7. Investment of organization in career*	Non-substitutable resources	0.742	0.740	0.000
8. Technological support to employees				
9. Willingness to achieve the goals at any cost.				
10. Employees add value	Resources add value	0.657	0.638	0.000
11. Organization culture and value*				
12. Employees are key for customer satisfaction				

Items with* are dropped items for analysis.

Table 3 Correlations between the items of psychological empowerment and competitive advantage

	CA	GI	PCO	PCM	
CA	1	.658**	.545**	.648**	CA= competitive advantage
GI		1	0.542**	.598**	GI= goal internalization
PCO			1	.665**	PCM= Perceived competence
PCM				1	PCO= Perceived control

To find the impact of variables after the extraction of factors, regression analysis was undertaken between the variables. Regression analysis was used in order to compare and analyze the relationship of the variables. To test the significance of the result, Ttest, F test and R^2 test has been carried out.

The data collected from the respondent are presented, analyzed and interpreted for attaining the objective stated in the study as follows:

10. Regression analysis:

After the extraction of the different variables and correlations between different variables is calculated, regression is done to find out the relationship between competitive advantage and psychological empowerment. Regression between above stated variables is calculated to find out the impact of psychological empowerment in competitive advantage. Competitive advantage is calculated after the sum of its variables; resources add value, resources are in-imitable, resources are rare and resources are non-substitutable. Competitive advantage is taken as the dependent variable and the variables of psychological empowerment are taken as independent variables. Step wise regression is done to find out the impact. Variables of psychological empowerment are perceived competence, perceived control and goal internalization. Model for regression is stated in the following ways:

Table 4　Regression table

Model Summary			
	R	R Square	Adjusted R Square	Std. Error of the Estimate
1	0.684a	0.468	0.456	0.93850
2	0.795b	0.632	0.619	0.98676
3	0.816c	0.665	0.648	0.89420

$Y = b_0 + b_1x_1 + b_2x_2 + \cdots + b_kx_k$
Competitive advantage= $b_0 + b_1$ GI+ b_2 PCM +b_3 PCO
a. Predictors: (constant), GI
b. Predictors: (constant), GI, PCM
c. Predictors: (constant), GI, PCM, PCO
d. Dependent variable: competitive advantage- CA

Here,
PCO = Perceived control
GI = Goal internalization
PCM = Perceived competence
CA = competitive advantage

Value of R, R^2 and adjusted R^2 is noted to be increasing as the researcher goes on increasing the predictors as the independent variables. After entering all the variables value of R, R^2 as well as adjusted R^2 are greater than 0.6. This tells us that the independent variables are capable in defining the dependent variable which is competitive advantage.

Table 5 Anova table

		ANOVA			
	Sum of Squares	df	Mean Square	F	Sig.
Regression	1658.574	3	552.858	72.786	0
Residual	835.52224	110	7.5957		
Total	2494.09624	113			

Above Anova table shows that variables of psychological empowerment and structural empowerment are defining the dependent variable competitive advantage significantly as the Value of F is high and is significant as the p value is 0.000.

Hypothesis1: There is impact of psychological empowerment in competitive advantage is accepted at 1 percent level of significance.

Table 6 Coefficient table with Collinearity statistics

	Unstandardized coefficients		Standardized coefficients			Collinearity statistics	
	B	Standard error	Beta	t	sig	Tolerance	VIF
Constant	0.889	0.333		9.255	0		
GI	0.121	0.078	0.804	8.457	0	0.532	1.879
PCM	0.087	0.081	0.585	6.237	0	0.368	2.717
PCO	0.089	0.068	0.634	7.756	0	0.335	2.9851

All the three variables have shown the impact on the dependent variable which is competitive advantage as follows:

· Every additional one percentage increase in goal internalization is responsible for increasing twelve percentage increases in competitive advantage of the organization, provided the effects of other variables are held constant. This shows that if the employees perceive that the organizational goals achievement will lead towards the fulfillment of their goals or interest; this will increase the competitive advantage of the organization. This also shows that psychological empowerment has the impact on the competitive advantage.

· Every additional one percentage increase in perceived competence is responsible for increasing nearly nine percentage increases in competitive advantage of the organization, provided the effects of other variables are held constant. If the employees perceive that they are competent in doing their works, competitive advantage for organizations can be created.

・Every additional one percentage increase in perceived control is responsible for increasing eight percentage increases in competitive advantage of the organization, provided the effects of other variables are held constant. If the employees think that they have authority to make decisions and can influence decision made in the organization as well as influence the process in which works are done, competitive advantage can be increased.

・Impact of the variable goal internalization is the highest one where as the perceived control is the lowest one. Impact of other variables is seemed to be medium.

・There is no problem of multi-collinearity in the model since VIF values are close to unity.

Either the model fulfills its assumptions or not, different other tests are done by the researcher which is discussed below:

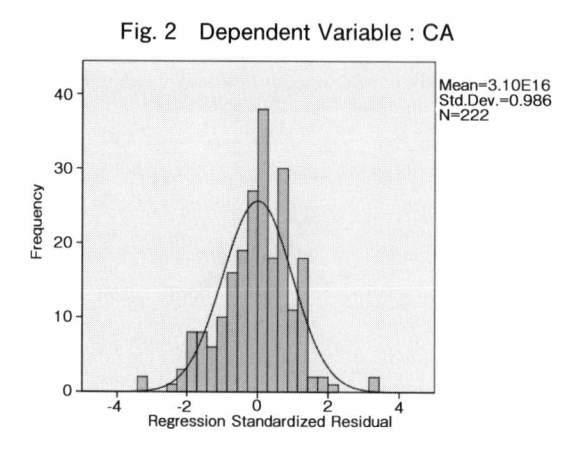

Fig. 2 Dependent Variable : CA

Histogram exhibits that residuals are normally distributed.

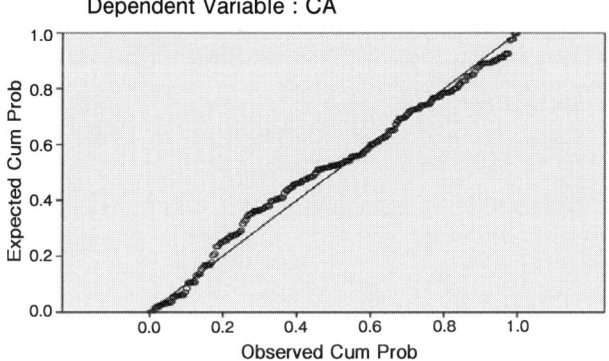

Fig. 3 Normal p-p plot of Regression Standardized Residual
Dependent Variable : CA

Normal p-p plot also exhibits that the residuals are normally distributed.

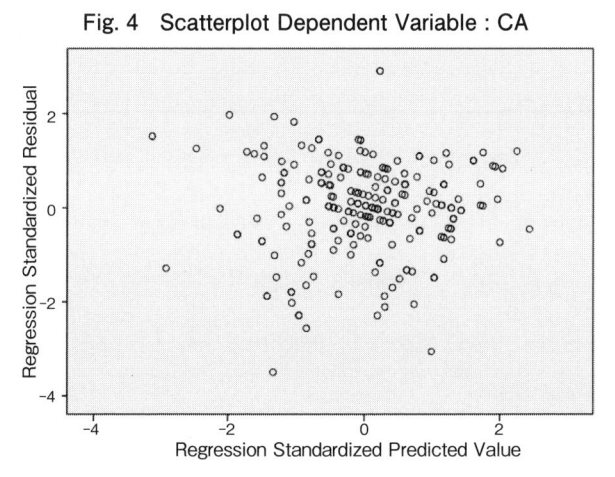

Fig. 4 Scatterplot Dependent Variable : CA

The null plot indicates random pattern. There is no problem of heteroscedasticity.

Table 7 One-Sample Kolmogorov-Smirnov Test

		Unstandardized Residual
N		222
Normal Parameters[a,b]	Mean	.0000000
	Std. Deviation	0.89078768
Most Extreme Differences	Absolute	.068
	Positive	.041
	Negative	−.068
Kolmogorov-Smirnov Z		0.78
Asymp. Sig. (2-tailed)		.263

One sample K-s test shows that the error terms are normally distributed since the p-value of the test is 0.263> 0.05.

11. Test of other hypotheses

For the test of other hypotheses t- test, Anova test and chi-square test are carried out which are discussed as below:

Hypothesis 2a: Across the two types of organizations, there is difference in organizational mean competitive advantage.

Table 8 t-test

Group Statistics			
	Mean	Std. Error Mean	N
Public hospital	210	3.44194	54
Private hospital	220.36	3.76248	60
absolute difference	10.36	5.52	

t=1.876, p=0.037

As the p value is 0.037, it is significant at 5 percent level of significance. So, we can conclude that there is difference in the competitive advantage between public hospital and private hospital.

Hypothesis 2b: Across the two types of organizations, there is difference in organizational mean psychological empowerment.

Table 9   t-test

Group Statistics				
	Mean	Std. Error Mean	N
Public hospital	35.2949	1.0234	54
Private hospital	38.4167	1.1540	60
absolute difference	3.12179	1.75397	

t=1.924, p=0.047

As the p value is 0.047, it is significant at 5 percent level of significance. So, we can conclude that there is difference in the psychological empowerment between public hospital and private hospital.

Hypothesis 3a: Across the different ages of organizations, there is at least one type of organizational mean market share coverage that significantly differs from others.

Table 10   ANOVA- market share coverage

	Sum of squares	df	Mean square	F	sig
Between Groups	12.680	2	6.340	11.236	0.000
Within Groups	123.832	220	0.563		
Total	136.511	222			

As the F value is significant, there is difference on market share coverage on the basis of age of the organization. To know which one age group has the highest post hoc analysis is carried out.

Table 11   Post hoc analysis on market share coverage

Age of organization	Mean difference	Standard error	Significance
Below 10 years and 10-50 years	0.57964*	0.12808	0.000
Below 10 years and above 51 years	0.08730*	0.11683	0.456
10-51 years and above 51 years	0.49234*	0.12905	0.000

From above table, it is seen than organizations in the age 0f 11-50 have the highest market share coverage.

Hypothesis 3b: Across the different ages of organizations, there is at least one type of organizational mean competitive position in terms of sales that significantly differs from others.

Table 12 ANOVA- competitive position of organization in terms of sales

	Sum of squares	df	Mean square	F	Sig
Between Groups	12.251	2	6.126	13.284	0.000
Within Groups	101.444	220	0.461		
Total	113.695	222			

As the p value is 0, tested hypothesis is accepted at 5 percent level of significance i.e. there is mean difference in competitive position of organizations in terms of sales between the organizations on the basis of their age. To know the actual situation post hoc analysis is done whose result is given below:

Table 13 Post hoc analysis on competitive position in terms of sales

Age of organization	Mean difference	Standard error	Significance
Below 10 years and 10-50 years	0.49261*	0.11593	0.000
Below 10 years and above 51 years	0.47707*	0.10575	0.000
10-51 years and above 51 years	0.01554	0.11680	0.894

After the post hoc analysis, highest mean of trend of competitive position in terms of sales is seen in the middle aged organizations where lowest is seen in the newly established organizations. Old organizations and middle aged organizations competitive position in terms of sales seems to be significantly different from newly established organizations.

Hypothesis 4a: There is significant difference in mean for competitive advantage between married and unmarried.

Table 14 t-test for competitive advantage on the basis of marital status

	Mean	Standard error	N
Married	31.1563	0.70013	64
Unmarried	31.9241	0.35626	53
Absolute difference	0.76789	0.78556	

t= 0.977, p= 0.331

As the p value is high, there is evidence to reject tested hypothesis; i.e. there is no significant difference in mean for competitive advantage in between married and unmarried.

Hypothesis 4b: There is significant difference in mean for competitive advantage between male and female.

Table 15 t-test for competitive advantage on the basis of gender

	Mean	Standard error	N
Male	31.4537	0.4444	48
Female	32.0446	0.46001	56
Absolute difference	0.59094	0.63965	

t= 0.924, p= 0.357

As the p value is high, there is evidence to reject tested hypothesis; i.e. there is no significant difference in mean for competitive advantage in between male and female.

12. Major Findings with Discussions

Competitive advantage and empowerment are two sides of a coin and are highly popular in the contemporary literature of management nowadays. The reasons behind such popularity include the rapid change that organizations face today, the complexity of the business environment, the impacts of globalization and unstructured markets, the ever changing consumer needs, competition, the revolution of information technology and communications, as well as the liberation of global trade.

Regression analysis has given the significant result with the conclusion that empowerment has the impact on the competitive advantage. Psychological empowerment has positive and significant impact on the competitive advantage of an organization. Anova test also showed the significant result. Tolerance and VIF test has shown absence of multicollinearity as TOL values are less than 1 whereas value of VIF are less than 10. Different other plots and tests discussed in the analysis part shows that data are normally distributed and there is absence of heterocsedasticity.

In this research also, three dimensions of Psychological empowerment as that of Menon are extracted as goal internalization, perceived competence and perceived control which all has the impact on the competitive advantage. From the standpoint of researchers interested in competitive advantage, the resource based view of the firm provides a framework for examining the role of human resources in competitive success and forces us to think more clearly about the quality of the workforce skills at various levels and the quality of the motivation climate created by strategic human resource management (Boxall, 1996). This research also has shown that Psychological empowerment has the impact on the competitive advantage of the organization. So, organizations should give emphasis on the empowerment of their employees psychologically so that organizational competitive advantage could be increased.

As positive relation is found in this study between the Psychological empowerment and Competitive advantage, Hospitals should focus on development and implementation of those HR policies which can increase the degree of empowerment leading to competitive advantage of the organization. Research carried out by Yukl et al. (2006) identified meaningfulness, competence, choice and impact as the four variables of psychological empowerment during their research but Nepalese findings are to some extent different than their studies in terms of variables. They found that in order to be sustained, empowerment needs to be a part of the long term strategy of the organization. Empowerment initiatives should be guided by the dual objectives of improving organizational effectiveness and improving the quality of work-life for employees which is not exception in context to Nepal.

On the basis of the age of the organization, it is found that hospitals between ages of 11-50 are more effective in organizational market coverage as well as in competitive position in terms of sales. No difference is found in competitive advantage between male and female as well as married and un-married. From this we can say that hospitals in middle age shows high degree of empowerment and competitive advantage. Organizations at the starting point can be struggling so not showing high degree of empowerment and Competitive advantage whereas organizations at higher ages may need innovative HR practices so the degree of both the empowerment and Competitive advantage could be increased.

Significant difference is found in the competitive advantage between the public sector and the private sector hospitals. Public hospitals are seen the most poor in increasing their competitive advantage through the resource based view of the organization. Public hospitals have to work more on the implementation of innovative HR practices for upgrading their quality for long term survival. Gender and marital status show no link with the competitive advantage of the organization. So, despite of marital status and gender organizations should focus on empowering all by providing equal opportunities.

From research studies of Spreitzer (1992), Thomas & Velthouse (1990), four aspects of empowerment have been conceptualized as cognitive components of intrinsic task motivation. These four aspects of empowerment are: (a) autonomy, defined as a sense of freedom in making choices about how to do one's work, and the resulting feelings of personal responsibility for these choices; (b) competence, defined as the belief in one's ability to perform a job successfully; (c) meaningfulness, defined as the perceived value of one's job in relation to one's personal beliefs, attitudes, and values; and (d) impact, defined as the belief that one is producing intended effects and has control over desired outcomes through one's task behavior. Menon (2001) found three dimension of psychological empowerment as perceived control, perceived competence and goal internalization. The dimension of perceived control captures the effects of traditional empowerment techniques, such as delegation, increased employee autonomy and so on. The dimension of perceived competence has parallels in Conger and Kanungo (1988) model, the Thomas and Velthouse (1990) model, the Spreitzer (1995) model. It is a reflection of commitment to organizational objectives and goals that transformational leaders want to engender in their employees (Menon, 2001). The dimension goal internalization is another dimension of psychological empowerment. In this findings, goal internalization, perceived control and perceived competence are identified as the dimensions of psychological empowerment. That's why, it can be suggested to the Hospital sector of Nepal that the three dimensions of psychological empowerment seems to be adopted to increase competitive advantages.

This research has also shown that empowered human resources are capable in increasing the competitive advantage of an organization. It is essential that the hospitals should focus on empowering employees psychologically so that competitive advantage could be increased. This research also shows that empowered human resources can be a source of competitive advantage as suggested by Barney (1991), being R^2 significant in the model used in this research.

Boxall (1998), states that Human capital advantage is possible if firms employ people with valuable but rare knowledge and skills, which are embedded by being to some extent firm —specific. To achieve sustained advantage through people, Boxall argues that management must nurture resources and processes that bring about high mutuality with talented workers and must similarly invest in employee and team development. The findings of this study indicate that public hospitals are showing quiet low degree of employee empowerment as well as competitive advantage in comparison to private sectors. This can create a vicious cycle in the health of the middle level people as well as lower level people because the target or the customers of the public hospitals are these people. This does not only have adverse impact on the health of these people but also in the economy of this level. For the development of the nation as well as to ensure the long term sustainability of these organizations they have to focus on empowerment so that competitive advantage of these hospitals could be increased.

13. Conclusion

Hospitals in Nepal have followed the quite traditional technique for granting the authority. All power and authority are resumed by top level and decisions are made by them. But the lower level manager are those people, who are "close to action" and face problems while dealing with patients as well as employee, hence, they must be participated in decision making.

As this research has shown that empowered human resources are capable in increasing the competitive advantage of hospitals, it is essential that they should focus on empowering health workers psychologically so that competitive advantage of the hospitals could be increased. This conclusion is based on the result of regression analysis and Anova test.

This research as has shown that public hospitals are showing quiet low degree of employee empowerment as well as competitive advantage in comparison to private sectors. This can create a vicious cycle in the health of the middle level people as well as lower level people because the target of the public hospitals are these people. This does not only have adverse impact on the health of these people but also in the economy of this level. For the development of the nation as well as to ensure the long term sustainability of these hospitals they have to focus on empowerment so that competitive advantage of these hospitals could be increased.

I hope this research will be quiet supporting for researchers in future to explore other dimensions of competitive advantage and psychological empowerment so that they could know more about it which will ultimately result in the long term sustainability of the hospitals as well as the nation.

<References>

Adhikari, D. R. (2002) 'Human resource management: An agenda for the future research in Nepal', *Banijya Sansar*, No. 10, pp. 26-30.

—— and Gautam, D. K. (2011) 'Employees Commitment and Organizational Performance in Nepal: A Typological Framework', *SEBON Journal*, Vol. 1, No. 17.

—— (2012) 'Status of Corporate Social Responsibility in selected Nepalese Companies', *Corporate Governance: The international journal of business in society*, Vol. 12, No. 5.

Adinolfi, P. (2003) 'Total quality management in public health care: A study of Italian and Irish Hospitals7', *Total Quality Management*, Vol. 14, No. 2, pp. 141-150.

Alemu, K. T., Yosef, S. A., Lemma, Y. N.and Beyene, S. Z. (2011) 'Strengthening Public sector Human Resource Management Capacities in Africa', Retrieved from Unpan 044910. pdf, SHRMinhealthsector. pdf.

Argyris, C. (1998) ' Empowerment: The emperor's new clothes', *Harvard business review*, the May-June 1998, Issue, pp. 98-105.

Bach, S. (1994) 'Restructuring the personnel function: the case of NHS trusts', Human Resourcemanagement Journal, Vol. 5, No. 2, pp. 99-115.

Bae, J. and Lawler, J. J. (2000) 'Organizational and HRM strategies in Korea: Impact on firm performance in an emerging economy', *Academy of Management Journal*, Vol. 43, No. 3, pp. 502-517.

Baniya, L. B. (2004) 'Human Resource Development Practice in Nepalese Business Organizations: A Case Study of Manufacturing Enterprises in Pokhara', *The Journal of Nepalese Business studies*, Vol. 1, No 1.

Barnett, S., Buchanan, D., Patrickson, M. and Maddern, J. (1996) 'Negotiating the evolution of the HR function: practical advice from the health care sector', *Human Resource Management Journal*, Vol. 6, No. 4, pp. 18-37.

Barney, J. B. (1991) 'Firm Resources and Sustained Competitive Advantage', *Journal of Management*, Vol. 17, No. 1, pp. 99-120.

—— and Wright, P. M. (1998) 'On becoming a strategic partner: The role of human resources in gaining competitive advantage', *Human resource management*, Vol. 37, No. 1, pp. 31-46.

Bartram, T., Casimir, G. and Fraser, B. (2007) 'Lost in translation: exploring the link between HRM and performance in healthcare', *Human Resource Management*, Vol. 17, No. 1, pp. 21-41.

1 Psychological Empowerment for Competitive Advantage 243

Boxall, P. (1996) 'The Strategic Human Resource Management Debate and the Resource-Based View of the Firm', *Human Resource Management Journal*, Vol. 6, No. 3, pp. 59-75.

——— (1998) 'Achieving Competitive Advantage through Human Resource Strategy: Towards a Theory of Industry Dynamics', *Human Resource Management Review*, Vol. 8, No. 3, pp. 265-288.

Bowen, D. E. and Lawer, E. E. (1992) 'The Empowerment of Service Workers: What, Why, How and When?', *Sloan Management Review*, Spring 33, p. 31.

Baniya, L. B. (2004) 'Human Resource Development Practice in Nepalese Business Organizations: A Case Study of Manufacturing Enterprises in Pokhara', *The Journal of Nepalese Business studies* volume 1, No. 1.

Bhandari. S. (2013) 'Empowerment and competitiveness: A resource based approach to Human Resource Management', thesis Submitted to Department of Management, T.U. Kathmandu.

Capelli, P. and Singh, H. (1992) 'Intergrating SHRM and strategic management' in Lewin, D., Mitchell, O. S., & Sherer, P. D. (Ed.), *Research frontiers in industrial relations and human resources*, pp. 165-192, NY: Cornell University Press.

Conger, J. A. and Kanungo, R. N. (1988) 'The empowerment Process: Integrating theory and Practice7', *Academy of Management Review*, Vol. 13, No. 3, pp. 471-482.

Charles R., Gowen III Kathleen L., McFadden William J. Tallon (2006) 'On the centrality of strategic human resource management for healthcare quality results and competitive advantage', *Journal of Management Development*, Vol. 25, Issue 8, pp. 806-826.

Chung, C. E. E. (2011) 'Job Stress, Mentoring, Psychological Empowerment and Job Satisfaction Among Nursing Faculty', Available at: http://digitalscholarship.unlv.edu/thesesdissertations/1266/, Accessed February 26, 2013.

Dadashinasab, M., Sofian, S., Asgari, M. and Abbasi, M. (2012) 'The effect of intellectual capacity on performance: A study among Iranian automotive industry', *Journal of basic and applied scientific research*, Vol. 2, No. 11, pp. 11353-11360.

Farrell, A., Souchon, A. and Durden, G. (2001) 'Service encounter conceptualization: employee's service behaviors and customer's service quality perceptions', *Journal of marketing management*, Vol. 17, pp. 577-593.

Fitzgerald, L., Lilley, C., Ferlie, E., Addicott, R., McGivern, G., Buchanan, D., Baeza, J., Doyle, M. and Rashid, A. (2006) *Managing Change and Role Enactment in the Professionalize Organization*, SDO Board, London.

Forrester, R. (2000) 'Empowerment: Rejuvenating a potent idea', *Academy of Management Executive*, Vol. 14, Issue 3, pp. 67-79.

Gautam, D. K. (2015) 'Strategic integration of HRM for organizational performance: Nepalese reality', *South Asian Journal of Global Business Research*, Vol. 4, Issue 1, pp. 110-128.

Gowen, C. R., McFadden, K. L. and Tallon, W. J. (2006) 'On the centrality of strategic human resource management for healthcare quality results and competitive advantage', *The Journal of Management Development,* Vol. 25, No. 8, pp. 806-826.

Gratton, L. (1977) 'Human Resource Strategy', *People Management,* Vol. 3, No. 15, pp. 22-27.

Guest, D. (1990) 'Human Resource Management and the American Dream', *Journal of Management Studies,* Vol. 27, No. 1, pp. 377-397.

Hart, C. and Schlesinger, C. (1991) 'Total Quality management and human resource professional: Applying the Baldrige framework to human resources', *Human resource management,* Vol. 30, No. 49, pp. 433-454.

Hyde, P., Boaden, R., Cortvriend, P., Harris, C., Marchington, M., Pass, S., Sparrow, P. and Sibbald, B. (2006) *Improving Health through Human Resource Management: Mapping the Territory*, CIPD, London.

Kabene, S. M., Orchard, C., Howard, M. J., Soriano, M. A. and Leduc, R. (2006) 'The importance of human resource management in health care: A global context', *Human Resources for Health,* 2006, Vol. 4: 20.

Kahreh, S. M., Ahmadi H. and Hashemi A. (2011) 'Achieving Competitive Advantage Through Empowering Employees: An empirical study', *Far East Journal of psychology and Business,* Vol. 3, No. 2, pp. 26-37.

Kamoche, K. (1994) 'A critique and proposed reformulation of strategic human resource Management', *Human Resource Management Journal,* Vol. 4, No. 4, pp. 29-43.

Kanji, G. K. and Sa, P. M. (2003) 'Sustaining healthcare excellence through performance Measurement7', *Total Quality Management,* Vol. 14, No. 3, pp. 269-289.

Khan, T. M., Saboor, A., Khan, A. N. and Ali, I. (2011) 'Cannotation of Employees Empowerment Emerging Challenges', *European Journal of Social Sciences,* Vol. 22, No.4, pp. 556-569.

Kraimer, M. L., Seibert, S. E. & Linden, R. C. (1999) 'Psychological empowerment as a multidimensional construct: A construct validity test', *Journal of management,* 59, pp. 127-142.

Lashley, C. (1995) 'Towards an understanding of employee empowerment in hospitability services', *Journal of contemporary hospitality management,* Vol. 7, No. 1, pp. 27-32.

Li, C., Wu, K., Johnson, D. E. and Wu, M. (2011) 'Moral leadership and psychological empowerment in China', *Journal of managerial psychology,* Vol. 27, Issue, pp. 90-108.

Marasini, B. R. (2003) 'New health policies and programs and involvement of non-governmental organizations health care (NGOS) -1951-1963. A health and hospital development in Nepal: Past and present', *Journal of Nepal Medical Association*, Vol. 42, pp. 306-331.

Menon, S. T. (2001) 'Employee Empowerment: An Integrative Psychological *Approach'*, *Applied Psychology: An International Review,* Vol. 50, No. 1, pp. 153-180.

National Health Strategy Consultative Forum (2001) 'National Health Strategy Consultative Forum Subgroup Report', *Service Delivery and HR, Department of Health and Children,* Dublin.

1 Psychological Empowerment for Competitive Advantage   245

Neilsen, E. (1986) 'Empowerment strategies: Balancing authority and responsibility', in S. Srivastva (Ed.), *Executive power,* pp. 78–110, San Francisco: Jossey-Bass.

Nunally, J. (1967) *Psychometric Theory*, New york, NY: Mc Graw Hill.

Operational research report (2012) *Human resources for health management form central to district level in Nepal Society* for local integrated development.

Pfeffer, J. (1994) *Competitive Advantage through People,* Harvard School Business Press, Boston MA.

——— (2005) 'Producing competitive advantage through the effective management of people', *Academy of management executive,* Vol. 19, No. 4, pp. 95–107.

Peteraf, M. (1993) 'The cornerstones of competitive advantage: A resource based view', *Strategic management journal,* Vol. 14, pp. 179–191.

Quinn, R. E. and Spreitzer, G. M. (1997) 'The road to empowerment: Seven questions every leader should consider', *Organizational dynamics,* Vol. 26, No. 2, pp. 37–49.

Randolph, W. A (1995) "Navigating the Journey to Empowerment", *Organization Dynamics,* Vol. 24, pp. 19–32.

Rawat, P. S. (2011) 'Effect of Psychological Empowerment on Commitment of Employees: An Empirical Study7', *International Conference on Humanities, Historical and Social Sciences,* Vol. 17, p. 143.

——— Report on census of private hospitals in Nepal, 2013. Government of Nepal, National planning commission secretariat; Central Bureau of Statistics.

——— Annual Report on department of health services, 2013/14. Government of Nepal, Ministry of Health and Population; Department of Health services.

Schuler, R. S. and MacMillan I. (1984) 'Gaining Competitive Advantage Through Human Resource Practices', *Human Resource Management,* Vol. 23, pp. 241–256.

——— and Jackson, S. E. (1987) 'Linking Competitive Strategies with Human Resource Management Practices', *Academy of Management Executives.* Vol.1, pp. 207–219.

Sparks, B., Bradely, G. and Callan, V. (1997) 'The impact of staff empowerment and communication style on customer evaluations: The special case of service failure', *Psychology and marketing journal,* Vol. 14, No. 3, pp. 475–493.

Shrestha, C. and Bhandari, R. R. (2012) 'Insight into Human Resources for health status in Nepal', *Health prospect,* Vol. 11, pp. 40–41.

Spreitzer, G. M. (1995) 'Psychological Empowerment in the Workplace: Construct, Definition, Measurement and Validation', *Academy of Management Journal,* Vol. 38, pp. 1442–1465.

——— (1996) 'Social structural characteristics of Psychological Empowerment', *Academy of Management Journal,* Vol.39 (2), pp. 483–504.

———, Kizilos, M. A. and Nason, S. W. (1997) 'A dimensional analysis of the relationship between psychological empowerment and effectiveness, satisfaction and strain', *Journal of management,* Vol.

23, No. 5, pp. 679-704.

Sundbo, J. (1999) 'Empowerment of employees in medium and small and me dium sized firms.', *Journal of Management,* Vol. 21, No. 2, 105-127.

Thamizhmanii, S. and Hasan, S. (2010) 'A review on an employee empowerment in TQM Practice', *Journal of Achievements in Materials and Manufacturing Engineering*, Vol.39, nr 2, pp.204-210.

Thomas, K. W. and Tymon, W. G. (1994) 'Does empowerment always work: Understanding the role of intrinsic motivation and personal interpretation', *Journal of Management Systems,* Vol. 6, No. 2, pp. 1-13.

―― and Velthouse, B. A. (1990) 'Cognitive Elements of Empowerment: An "Interpretive" Model of Intrinsic Task Motivation', *Academy of Management Review,* Vol. 15, pp. 666-681

Ulrich, D. (1991) 'Using Human Resources for Competitive Advantage', in R. Kilman, I. Kilman and Associates (Ed.), *Making organizations competitive,* pp. 129-155, San Francisco, CA: Jossey Bass.

―― and Lake, D. (1990) *Organizational Capability: Competing from the Inside Out,* New York, NY: John Wiley and Sons.

Van Beveren, J. (2003) 'Does health care for knowledge management?', *Journal of Knowledge Management,* Vol. 7, No. 1, pp. 90-95.

Wernerfert, B. (1984) 'A resource based view of the firm', *Strategic management journal,* Vol. 5, pp. 171-182.

West, M. A., Guthrie, J. P., Dawson, J. F., Borrill, C. S. and Carter, M. (2006) 'Reducing patient mortality in hospitals: The role of human resource management', *Organizational Behavior,* Vol. 27, No. 7, pp. 983-1002.

Wiig, K. M. (2002) 'Knowledge management in public administration', *Journal of Knowledge Management,* Vol. 6, No. 3, pp. 224-239.

Wright, P. and McMahan, G. (1992) 'Theoretical perspectives for strategic human resource management', *Journal of Management,* Vol. 5, pp. 301-326.

――, ―― and McWilliams, A. (1994) 'Human Resources and Sustained Competitive Advantage: A Resource-Based Perspective', *International Journal of Human Resource Management,* Vol. 5, pp. 301-326.

World Health Organisation. WHO Global Atlas of the Health workforce, Available at: http://www.who.int/globalatlas, Accessed January 13, 2015.

Yavas, U. and Romanova, N. (2005) 'Assessing performance of multi-hospitals organizations: a measurement approach', *International Journal of Health Care Quality Assurance,* Vol. 18, No. 3, pp. 193-203.

Yukl, G. A. and Becker, S. W. (2006) 'Effective Empowerment in Organizations', *Organization Management Journal, Linking theory and Practice: EAM White Papers Series,* Vol. 3, No.3, pp. 210-

231.
Yung, C. C. and Lin, C. Y. Y. (2009) 'Does intellectual capital mediate the relationship between HRM and organizational performance? Perspective of a healthcare industry in Taiwan', *The international journal of Human Resource Management,* Vol. 20, No. 9, pp. 1965-1984.

Zairi, M. (1998) 'Building human resource capability in health care: A global analysis of best practice — Part III', *Health Manpower Management,* Vol. 24, No. 5, pp. 166-169.

2 高齢化地域の持続可能性に資する地域企業のイノベーション戦略
――徳島県上勝町「いろどり」からの考察

芳賀和恵

ドイツ日本研究所経営・経済領域専任研究員

【要旨】
高齢化が進む中山間地域は，従来，地域産業停滞，高齢化と人口減少の問題を抱え，地域維持がさらに困難になっていることが多い。その問題を解決する地域発企業の事業の可能性について，地域の自力で起こされるイノベーションに注目し，また，ルーマンのシステム論を参考にして考察する。

徳島県上勝町「彩」事業の事例から三点の含意が得られた。第一に，地域の優位性を活かし，他地域との差別化に成功すれば，ローテクの財が中山間地域の戦略になり得る。第二に，一般の事業会社が同時に地域の持続性にも資することは可能である。第三に，「彩」の生成プロセスはオートポイエーシス的である。

キーワード：持続可能性，地域経済，中山間地域，高齢化，イノベーション，アントレプレナーシップ，キーコンピテンシー，オートポイエーシス

1. 課題と背景

高齢化は先進国，さらに中国等の新興国が直面している問題である。日本は高齢化が最も進行していることに加えて，高齢化が急激に進行したことにより，高齢化先進国としてその対応が各国からの関心を集めている。日本の高齢化は，中山間地域において進行度が高い。高齢化による問題は多岐に渡る。

本稿では高齢化の問題を抱える中山間地域の地域経済について論ずる。人口流出による人口減少，高齢化と経済力の低さを，相互に関連する問題ととらえ

て，自立性と持続性を持つ地域経済の可能性を探るものである．

　企業の経済的パフォーマンスのみを目的とした経営資源や能力の最適配分をするだけでは，長期的に持続可能な経済発展は実現できない．企業活動が企業の利益だけではなく，社会的持続性にも資することは不可避となっている．地域の経済力の構造的な弱さと高齢化および人口減が，別個の問題ではなく，相互に関連する問題であると考えるならば，地域経済活性化によって相互関連の関係にある高齢化，人口減少，さらにそれによって引き起こされる社会的問題が解決され得ると考えられるであろう．そのために地域の企業に求められることは，企業の事業内容や，組織の構成，マネジメントなどを，地域の持続可能性に適するようにデザインすることであろう．そのような，地域にとって包括的な効果をもたらす事業とは，具体的にどのような事業内容であろうか．企業の経済的パフォーマンスが，コミュニティを犠牲にすることなく，むしろそれらの質を向上させる，社会問題解決的な価値創造をする事業はどのような要素によって構成されるのだろうか．地域の持続性維持の問題を抱える中山間地域の住民が自分の力をもってそのような事業をおこすことは，いかにすれば可能であろうか．

　具体例として，徳島県上勝町の「彩」事業を考察する．「彩」事業は，中山間地域再生の成功事例として注目を集めている．そして，サービス受給者と考えられがちな高齢者が「彩」生産の中心となっていること，事業を通じて高齢農家が生きがいを強く感じるようになったことで評価が高い．また，「つまもの」を市場化したことで，事業内容に独自性が高い．一般の事業会社の営利追求の事業活動が，地域の抱える社会的，経済的問題も同時に解決するあり方の1つを具現していると思われる．その発展プロセスについて，自己構成的性質に焦点を当てて考察する．

2. 中山間地域の抱える問題と政策

2-1. 地域経済の歴史的経緯

　従来，工業地帯の発展を経済発展の中心に据えた第二次世界大戦後の日本の

経済政策により，地方は工業地帯への労働力供給源として工業地帯発展の枠内での機能を果たしてきた。平行して，第一次産業の競争力の相対的，絶対的な低下が起こった。戦後の復興過程から高度成長期前半まで，わが国では経済発展が最優先の課題とされ，既存の臨海工業地帯の基盤整備に重点が置かれていた（経済産業省, 2004）。その結果，農村部から都市部への人口流入が増大し，結果として過疎化の問題が生じた。

地域経済は「まちづくり」の一部であると考えられる。石川公彦（2013）は，地域の持続性を形成することを2つの次元で考えている。第一の方向性は，地域経済の活性化を目指す取り組みでる。第二の方向性は，住空間の向上運動等を含む物理的防衛，あるいは，生きがいをつくり，文化的な営みを充実させるような生活向上運動などを含む取り組みである。

ルーマンのシステム論によれば，社会的なシステムがより厳密なカップリングの状態に入れば入るほど，当のシステムはより高いリスクを負うことになり，実際にもより大きな損害に見舞われることになる（Luhmann, 2002）。したがってシステム同士は緩いカップリングの時に，環境の変化や複雑性に柔軟に対応できるであろう。経済，行政などのシステムは独立性を保ち，ハーバート・サイモンの「限定合理性」により情報や条件を限定することによって，おのおのの専門領域に集中することになる（Luhmann, 2002）。そうでありながらも全体としての安定性が得られるためには，経済，行政その他のシステムが，それぞれ独自の領域に特化しながらも，他のシステムと結果的に調和的にコミュニケーションし続けるということである。

中央の政策では，地域が主体となる地域の自立のコンセプトは，すでに謳われている。1998年に策定された地域の自立の促進と美しい国土の創造を目指した第五次の全国総合開発計画である「21世紀の国土のグランドデザイン」では，「自立の促進と誇りの持てる地域の創造」をはじめとした5つの基本課題を掲げるとともに，国土基盤投資について地域の特性を十分に踏まえた投資，次の世代に備えた投資の重要性，産業政策面では地域の産業競争力を再生することで地域の自立を促す点が指摘された（経済産業省, 2004）。さらに，地域の視点で地域が直面するさまざまな課題の解決を推進するため，2003年には地

域再生本部が発足し,地域経済の活性化と地域雇用の創造の実現のためには,地域の産業,技術,人材,観光資源,自然環境,文化,歴史など地域が有するさまざまな資源や強みを知恵と工夫により有効活用し,文化的・社会的つながりによる地域コミュニティの活性化や,民間事業者の健全なビジネス展開を通じ十分な雇用を確保することが重要であるとしている(経済産業省,2004)。しかし,これまで中央集権的な色彩が強く,地方自身において包括的な地域経済政策立案を行う機会がなかったため,思うように成果が上がらない。

山村地域に対しては,1965年に山村振興法の制定により,地域経済強化,新しい林業管理や農業のシステム,農産物の加工産業の振興による山間地域の雇用機会の増加の促進が図られた。農水産業の付加価値をつけ,経済力を強化しようとする第六次産業化や農業クラスターも政策として提唱された。第六次産業化は,農産物を加工,レストランやテーマパーク設置などの展開によってさらに消費者へ接近することにより,差別化や品質管理の優位性を獲得しようとするものである(斎藤,2012)。これは,地域住民サイドにとっては,小規模な直売所やレストランでの高齢者や女性を含めた地域雇用の拡大によって地域全体の所得の拡大を目指すものである(斎藤,2012)。農業クラスターは,第六次産業化が点としての活動であるのに対して,地域という広がりに展開された農産物の付加価値活動を意味することが多い。どちらの場合も,個々の企業が目指すところは,農産物のブランド化である。しかしながら,成功事例が上がらず,問題となっている(Feldhoff, 2013;斎藤,2012;櫻井,2010)。

2-2. 農業クラスター,第六次産業の問題点

ハーシュマン(1961)はすでに1958年に産業間の連関による経済発展を主張した。すなわち,前方あるいは後方連関,ひいては前方後方連関によって,地域の産業が他の産業にリンクすることで,需要が生まれ,地域の収入増となる。ハーシュマンが指摘したように,農産物はそれ自体の価値が低い。また,中村(2014)は,技術が進歩して産業の高度化がもたらされたものの,地方都市では,地域内での産業間の関連構造が希薄であったことで,技術進歩の便益を享受した内生的発展に困難があったと指摘する。これにより,さらに農産物

の価値が上がらなかったと推察する。農業クラスターや第六次産業化は，他の産業への連関をそれぞれの形で実現するコンセプトだと考えられる。しかしながら，高齢化の進む中山間地域では，地域企業がこれらの政策に沿って活動することは難点があると考えられる。

　農業クラスター政策は，ポーターの産業クラスター政策をベースとしている。クラスター政策は，知識を集約して一地域がイノベーションが起きやすいようにするイノベーション戦略である（Krugman, 1991）。最新知識集約型のイノベーションの立地として，大学等の知の拠点を持たない中山間地域はクラスター戦略において比較優位性を有するとは言い難い。Smallbone and North (1999) も地域の中小企業はイノベーションを起こす能力が低く，新規技術の導入も遅いと指摘している。他拠点にある大学等からの知識移転は可能である。しかし，持続的成果を上げるためには，知識移転に時間がかかる。

　第六次産業化は，農作物を工業，商業などの農業の川下分野につなげるものである。三重県伊賀市の「伊賀の里モクモク手づくりファーム」が第六次産業化の成功事例としてしばしば挙げられるように（斎藤, 2012；関, 2014），主として，既存の農作物をブランド化してレストラン展開する例が多い。しかしながら，第六次産業化は，農業の川下領域との連携作りの難しさ[1]と，テーマパーク開設などの統合化のための投資額の大きさにより，普及が進まなかった（斎藤, 2012；櫻井, 2010）。高齢者や女性を含む地域での雇用拡大の点でも，本格的にレストランを成功させるためには，人材の質の向上が必要になり，質の高い人材の確保が課題となる（斎藤, 2012）。また，高齢化率が高く，高齢農家が農業の主たる担い手となっている中山間地域においては，一般の農作物において高齢農家は若手農家に対して体力的に劣位である。比較優位性が劣る商品（農作物）を，地域活性化策の中心にすることは，効果的な戦略ではないと思われる。

3. 中山間地域の持続性に資する事業

3-1. 農業製品ブランド化の商品領域

　自立できる経済システムとは，「地域が自らの生活の糧を稼ぎ出せること」

(中村,2014)を可能とする経済システムであり,1つには自らが域外マネーを獲得し,もう1つには域内で所得(付加価値)を生み出す(中村,2014)。経済効果とは,1年間の生産活動に派生してどのくらい各産業への生産需要が生まれ,最終的に地域全体の生産額がどのくらい増えるかという問題である。

前節で,地方の地域は技術に(比較)優位性がないと述べた。この点から,中山間地域においても,中村(2014)が指摘するような,地域の比較優位性を生かし,インパクトの高い価値を持つ新規市場を目指す事業が効果的だと考える。多くの場合は新規技術が財の開発の中心的な要素とはならないであろう。

革新性と技術力の組み合わせから新規市場の可能性を考えると,次のようなマトリックスであらわせるであろう(図1)。図中の技術的革新性が低く,イノベーションのインパクトが高い第二象限が,中山間地域に適するだろう。Kim and Mauborgne(2005)は価値と革新が等しく重んじられるイノベーションを企業の戦略として着目し,バリューイノベーションとよぶ。バリューイノベーションの達成目標は,新規市場「ブルー・オーシャン」である。財を受け取る側にとって魅力的な新価値を生成するバリューイノベーションが,ブルー・オーシャン創生の不可欠な要素であるとしている。Kim and

図1　インパクトと技術革新性によるイノベーションの類型

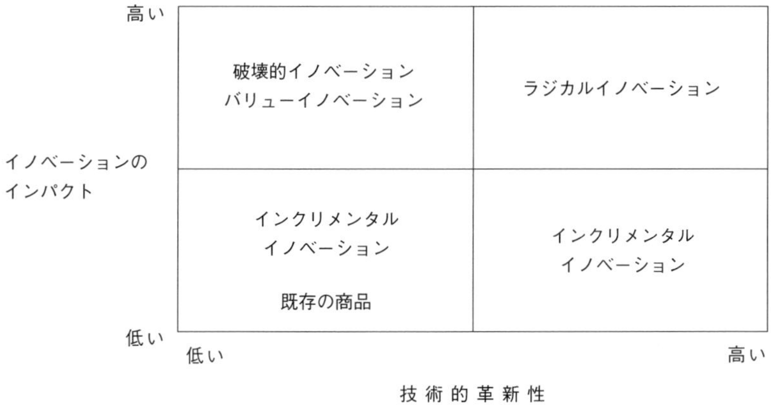

	技術的革新性 低い	技術的革新性 高い
イノベーションのインパクト 高い	破壊的イノベーション バリューイノベーション	ラジカルイノベーション
イノベーションのインパクト 低い	インクリメンタルイノベーション 既存の商品	インクリメンタルイノベーション

出典:Haga, 2013に基づきマトリックスを作成。

Mauborgne（2005）によれば，ブルー・オーシャンを生み出す企業は，規模，業界，企業本拠地などの属性がさまざまな領域で見られる。中山間地域においても，ブルー・オーシャン戦略による新規市場開拓が有効であると考える。

　Kim and Mauborgne（2005）の，需要側にとっての価値に着目し，競争をしないことを目指すブルー・オーシャンのコンセプトは，地域経済を考える上で有益であると考える。Kim and Mauborgne（2005）に倣えば，本稿で対象としている地域では，ローテクでインパクトの高い魅力的な新価値を作り出す分野への注力が有効であろう。Christensen（2000）が示した，既存市場での破壊的イノベーションによる競合他社との競争を展開するよりも，ブルー・オーシャン的な新規市場の創出がより適していると思われる。バリューイノベーションとしての付加価値をつけるために，新しい技術を活用することと，生産者の能力向上のための教育は大切である。これは，中山間地域が取り組む事業を他の類似の条件の地域の製品と差別化し，新たな価値が明確に認識される新規市場として位置付けるために必要であろう。

　Kim and Mauborgne（2005）は，バリューイノベーションは「新結合」(Schumpeter, 1993) を前提とせず，既存の財の価値の書き換えのみが必要だと述べている。しかしながら，農産物の現状を省みると，農業地域のバリューイノベーションは，農産物の既存の状態を保持したまま価値の書き換えを行うだけでは不十分であると思われる[2]。

　ブルー・オーシャンはイノベーションであり，新しい市場の創生であるので，ここで新市場の類型を考える。Röpke（2002）はニーズの有無と財の供給について市場の類型を考察している（図2）。既存の市場の周りに新規市場の可能性が存在し，三種類の類型が考えられる。地域にとって実現しやすいのは，領域1の「ニーズが顕在化／財は供給されていない」ケースか，領域2の「ニーズが顕在化していない／財は供給されている」ケースであろう。領域3の「ニーズが顕在化していない／財が供給されていない」ケースはRöpke（2002）によれば，ラジカルなイノベーションによって市場が作られる。農業クラスター政策の問題点から考えて，この領域には中山間地域が強みを持たないと考えてよいであろう。

図2　需要と供給の状況による市場の類型

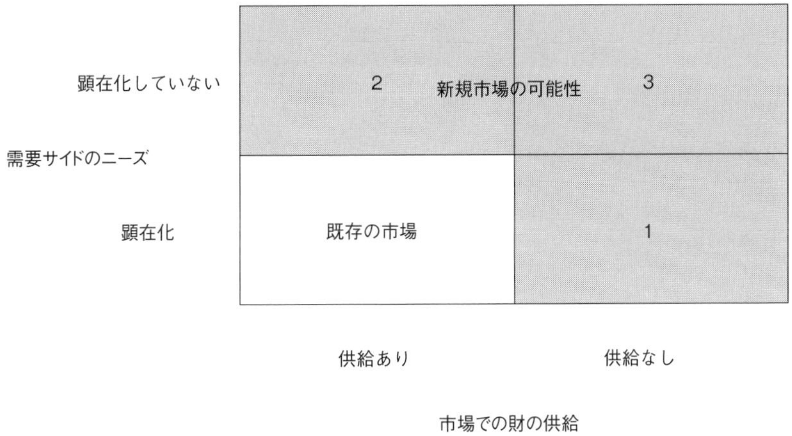

	供給あり	供給なし
需要サイドのニーズ　顕在化していない	2　新規市場の可能性	3
需要サイドのニーズ　顕在化	既存の市場	1

市場での財の供給

出典：Röpke, 2002, p. 167 を翻訳。

　地域の活性化のための経済おこしの手法は多様である。選択的な移住者誘致による地域活性化に成果を上げている徳島県神山町のように，外部の人材による活性化も手法の1つである。地域が地元の資源を有効活用し，地域住民が主体となる方法によって地域経済の活性化を目指す場合には，個々の企業が地域の比較優位性（立地条件，資源など）を活用し，地方住民の能力が効果的に用いられるような，インパクトの強いローテクな財を考え出し（開発し），Röpke (2002) の領域1あるいは領域2の性格を持つ市場を作ることが，事業の目指す戦略の1つであると考える[3]。

　地域の持続性に資する事業の内容や形態は，さまざまな組み合わせが可能である。中でも，地域住民から生まれた事業のアイディアが地域住民によって運営されるような事業の場合には，「21世紀の国土のグランドデザイン」や第六次産業化の農村地域活性化に謳われている高齢者や女性を含めた地元住民の積極的活用のコンセプトが実現するケースとして興味深い。高齢化率が50％を超える限界集落ともなれば，域外マネーの獲得をその集落の居住者に求めることは困難であるが，中村（2014）は，そういった場合にこそ集落とそれを取り

巻く自然環境と資源の中に経済的価値を認識することによって，行政にせよ民間にせよ，基盤産業を維持して地域の循環構造を維持していく取り組みが求められると主張している。経済の多様性が小さい地域経済の活性化の場合，往々にして他地域（自治体）とのリンクが必要となる。このために，ロジスティクスや通信分野での新規技術を取り入れることが必要であるケースが多いと想定できる。

地域住民が主体となるような事業では，事業内容の選択が地域住民が主体となれるような内容に限定されるという特有の制限を持つ。事例として取り上げる徳島県上勝町のように，高齢化率が高く，農業従事者が大半を占める場合には，高齢農家の健康状態，モチベーション，生活の質などが，事業における人的資本の質に関わる。地域住民の社会的満足度を高めることは，社会的企業の例に一般に見られるように，慈善的な社会的貢献を目指しているという理由にも基づき得るが，同時に事業成功ために達成すべき条件でもある。事業が社会的側面での成果を上げることへの二重の根拠を持つケースが多いであろう。さまざまな社会的企業のケース（谷本, 2009）の中でもユニークであると考える。

3-2. 組織：第三セクター

地域活性化の事業を行う組織として，しばしば第三セクターが選ばれる。この組織形態は持続的な地域経済に資するのであろうか。

日本における第三セクターの定義は，一定に定まっているとは言い難い（堀場・望月, 2007；入谷, 2008）。一般には，公共部門と民間部門の共同出資の形態の組織を指す。設立形式からは，さらに商法法人としての第三セクターと民法法人としての第三セクターに分かれる。商法法人としての第三セクターは，売上最大化を図る企業のように行動し，公共財ではなく私的財を供給する。ヨーロッパにおいては，共同組合，共済組合，アソシエーションといった相互扶助組織活動を含める活動領域として第三セクターが定義される。本稿における第三セクターは，商法法人としての性格を持ち，また組合組織的な相互扶助の性格を持つ組織を対象とする。共同組合的性格から見た第三セクターは，民間企業の効率性と公的サービスの社会的役割を結び付け得る組織である（堀場・望

月, 2007)。社会的性格の事業を考える上で，第三セクターのこの側面は意味がある。

1980年代後半以降，商法法人のうち株式会社の設立の増加が顕著であった。しかしながら，1997年以降は株式会社の設立数も第三セクター総設立数もともに減少傾向にある。堀場・望月（2007）によれば，地方公共団体の財政事情の悪化が設立数減少に大きな影響を与えている。また，バブル期にリゾート開発やビル開発など大きな投資を必要とする事業が第三セクターで企画された（入谷, 2008）ことも一因であろう。

中山間地域のような規模の小さい地域の場合には，そのような投資が大きい事業ではなく，組合組織的な性格を持ちながら小さい投資で地域固有の事業を選択するほうが有益であると考えられる。

3-3. イノベーションの発展プロセス

武石・青島・軽部（2012）は，イノベーションを「経済成果をもたらす革新」ととらえており，本稿でもこの解釈に倣う。イノベーションは，「革新」でなくてはならないが，単なる革新だけではなく「経済成果」をもたらさなくてはならない。すなわち，市場取引を通じて社会に経済的価値を提供するものである。

新しい事業（イノベーション）を起こすにあたっては，プロセスを通して企画をリードする機能が必要となる。シュンペーターはこの機能を「起業家」[4]とよんだ。その機能の本質は新結合（イノベーション）の実行にある。

プロセスにおいては，学習およびパートナー・協力者を獲得するネットワーク構築が大切である。これまで存在しなかったものを生成するイノベーションの実践のためには，「起業家」の学習プロセスが伴う（Röpke, 2002）。「起業家」自身にとって未知の事柄を事業対象とするため，それを可能にする学習が必要になるのである。この時，認知科学の知見にあるように学習者は既知の領域を足がかりに，類推できる領域へと学習を進めていく。上勝町の「彩」事業においても，その時点で利用可能な方法，手段を利用することによって成長してきた軌跡がみられる。

バリューイノベーションを創造するためには，現在より少し先の時点で価値が認められる事象をイメージする想像力と先見力（Hamel and Prahalad, 1994）が必要である。Karlsson（2012）は，この点に関して，地域経済の場合にも，ビジネス環境の変化から生ずる新たな可能性を素早く感知し，新規市場に育てる力が大切だと述べている。「起業家」が地元住民に対する共感，シンパシー，連帯感を持っている場合，より効果的なアイディアが生まれると思われる。着想を得てそれを育てる際にも，学習と情報を得るためのネットワーク構築がなされるであろう。

イノベーションを起こすには，他者の資源（ヒト，モノ，カネ，情報）が動員されなければならない。イノベーションの推進者（「起業家」）がなすべきことは，多様なルートを切り開き，組み合わせ，さまざまな支持者と，さまざまな理由を総動員して，壁を乗り越えて資源動員を創造的に正当化していくことである（武石・青島・軽部, 2012）。イノベーションは，今までにない初めてのことを実現しようとする革新である故に不確実性に満ちており，経済合理性を欠くために，他者の資源を動員するのが難しい。武石・青島・軽部（2012）は，事例の研究から，推進者が固有の理由で始動したイノベーションに対しては，やはり固有の理由から正当性を認める特定の他者と出会うことで，事業化へ向かう前進が可能になることを確認した。訴える相手に，不確実性のあるイノベーションの成功や効果を納得させるためには，相手に合った説得の仕方を選べなければならない。相手がどんな人物であるかを知り，相手の立場に自分を置いて考えられる能力も必要である。

学習とネットワーク形成が事業ごとに固有の経過をたどることは，現実にある例から見て取れる。しかしながら，それぞれの軌跡は，全くの偶然の結果ではないはずである。それぞれの自発性を説明するコンセプトは，研究の上でも，事業を実際に行うものにとっても，プロセスを整理する上で有益な手掛かりになると思われる。筆者は，ルーマンのシステム論的なアプローチが，このユニーク性を説明するために有益であると考える。生物学の視点である自己再生産的な生命システムに倣うマトゥラナとヴァレラのオートポイエーシス論，そしてそれをさらに社会システムに応用するルーマンのシステム論は，限界や批

判があるものの（谷本, 1993）[5], 次の点で本稿で扱うイノベーション生成プロセスには適していると考える。システムを固定的なものではなく，環境の変化に応じて組織構造が連続的に変化しながらも自己を維持し保存していくという，ルーマンのシステム論のコンセプトは，同じく連続的変化を前提とするイノベーションのプロセスに適していると考える。均衡状態を到達地点とせず，連続的変化を常態と考えるイノベーションプロセスをシステムとして捉える1つのモデルである。ルーマンは，社会システムを構成するオペレーションはコミュニケーションであると考える（Luhmann, 1984）。システムの諸要素は，一方では外界（環境）からの偶発性取り込みに対しては開かれているが，認識的には閉じられている。システムの構成要素が，環境からの刺激に対してどのように反応するかは，構成要素次第といえる。西口（2009）はルーマンのシステム論に拠りながら，社会システムにおける特定の個人や集団が行う意思決定や裁決による選択は自己準拠によってなされ，その一度なされた選択は，状況の変化に応じて，選択なされなかった選択可能性や新しく生じた選択可能性とともに再吟味されて再選択が起きることを指摘している。ルーマンのシステム論を参考にしているRöpkeのフィルターモデルによれば，人間の行動は「能力」，「モチベーション」と「行動の権利」の三要素により規定される（Combé, 2008）。個人それぞれに「能力」，「モチベーション」，「行動の権利」の領域があり，この三要素がすべて重なる部分が行動実行可能領域となり（図3），Schumpeter（2010）の「起業家のエネルギー」となるというものである。ルーマンのコミュニケーションにおける，外界からの刺激への反応を決める要素としても参考にできると考える。

イノベーションの実行者である「起業家」には

能力：イノベーションの着想を得る能力，アイディアを事業構想にする能力，実行能力，

モチベーション：アイディアを事業として企業活動として実践することへの欲求，

行動の権利：イノベーションを遂行する試行錯誤が可能な環境にあることが求められるであろう。

図3　フィルターモデル

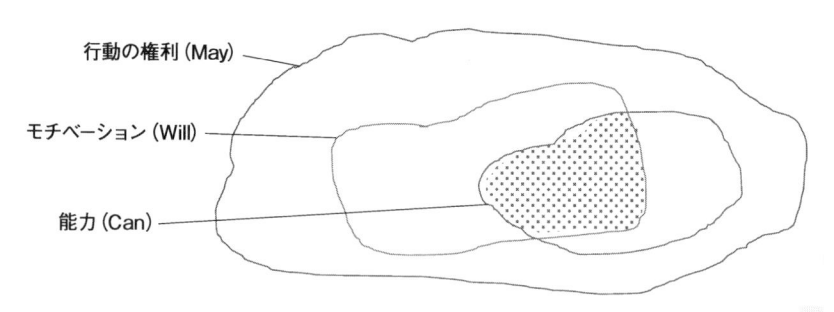

行動の権利 (May)
モチベーション (Will)
能力 (Can)

出典：Röpke & Xia, 2007, p.123より作成。

　Karlsson（2012）は，Bourdieu and Wacquant（1992）に拠り，持続性のあるネットワークは，成果物として事業に必要な要素であるソーシャルキャピタルを作りだすと主張する。ソーシャルキャピタルは，事業に必要な資源と品質を提供するものである（Karlsson, 2012）。実質的で持続性のあるネットワークが成立するためには，Westley, Zimmerman and Patton（2006）が関係性を強調するように，関係者全員が役割を果たし，関係性を育てていくようなさまざまな相互作用をネットワーク内に起こしていかなければならない。多様な関与者のどのような関わり方が望ましいかを考える必要があるであろう。また，西口（2009）は，ルーマンのシステム論を念頭に置きつつも，ネットワークの協力的な機能は，信頼の大きさが正の作用をおよぼし，信頼が大切だと述べる。「信頼」という概念が適当かどうかは議論の余地があるが，システムが自己準拠で環境からの刺激への反応を決定するにあたっては，なにがしかの要素が働いていると考えられる。

4. 事例「彩」[6]

　「彩」事業考察にあたっては，文献，インターネットからの情報，「いろどり」参与観察（2014年1月30日，31日）および上勝町役場での聞き取り（2014

年 1 月 31 日）の情報に基づく。

4-1．徳島県上勝町の概要

上勝町は徳島市内から約 40km，四国山脈の南東山地に位置する中山間地域である。総面積の 85.4％が山林で，標高 100m から 700m の間に大小 55 の集落が点在している。他の中山間地域同様，過疎と高齢化が同時進行している。人口は 1,767 名（2014 年 8 月現在），高齢者比率が 49.57％である（いろどり, n.d.）。上勝町の人口は 1970 年代から減少を続けている。現在の人口に対し 1950 年の人口は 6,356 名であった。

上勝町は，1950 年半ば以降，ミカンと木材の産地として発展してきたが，輸入木材による木材の低迷と，25 億円超の被害総額を出した 1981 年の寒波によるミカンの枯死により，一年間でミカンの売上は約半分となり，農業は大打撃を受けた（横石, 2007）。林業と農業が経済的比較優位性を失い始めてから，若年層の人口流出が始まった。

4-2．「彩」の分析

4-2-1．「彩」事業の概要と成果

「彩」事業は，料理の「つまもの」を市場商品化したものである。

かいしき（つまもの）市場の規模は 10 億円弱である。現在の「彩」事業の年商は 2 億 6,000 万円であり（図 4），かいしき市場の約 70％を占めている。

事業は，1986 年にスタートした。事業開始当時は 20 種類の製品ラインナップであったが，現在では 320 種類以上あり，一年を通してさまざまな葉っぱを出荷している（いろどり, n.d.）。発足時には 44 軒の農家が参加（横石, 2007），参加農家は順調に増え，2014 年 8 月現在，197 軒の農家が参加している。上勝町人口の約 10％が彩農家である。平均 70 歳，90％が女性である（笠松・佐藤, 2008）。参加農家は主に JA 東とくしま上勝支所の近くに住む農家である。つまものの注文は 10 時に発表されてから同日 13 時までに JA 東とくしま上勝支所に納品する。一覧表をファックスで受け取った農家は，自分がやりたい，出荷できる商品をすぐに農協に電話連絡する。誰がどの注文をとるかは，早い者

勝ちで決まる（横石, 2007）。農家は，自営の立場で「彩」事業に参加しており，収入はそれぞれの参加程度に応じて大きくばらついている。多い者では，年収1,000万円を稼ぐ農家もいる（いろどり, n.d.）。

「彩」事業の運営では，「つまもの」の標準化と供給の安定化を厳しく管理している。横石は「つまもの」を「彩」事業で商品化する過程で，商品となる葉の大きさ，色彩，外見，また葉っぱの意味，季節感を自ら学習し，市場で求められる精度と品質を農家に伝えた。これはイノベーションの大きなポイントだと考える。これによって品質の標準化がなされたことで，誰でもつまものを生産でき，市場商品として大量生産できるようになったからである。さらに，「彩」事業では，つまもの商品の出荷の前に料亭の注文をあらかじめ確定するシステム[7]を確立している。価格の安定のためと市場の不確かさからくる不安を仲卸業者から取り除くことにより，販売に寄与したといえる（石川和男, 2013）。また，「彩」事業が全国に市場を持てたのは，「彩」事業では既存の農業技術のみならず，コンピュータや情報技術を積極的に取り入れていることに

図4　「彩」ブランド売上高推移

出典：いろどり。
注1：「彩」商品の販売はJA東とくしま上勝支所の担当であり，売上はJA東とくしま上勝支所のものである。
注2：図中の売上高は，「彩」商品の一部分である。

よる。一例をあげれば，2006年に上勝町全町に光ファイバーが敷設され，イントラネットで結ばれていたいろどりシステムネットワークがインターネットで結ばれた（笠松・佐藤，2008）。この点で，他の中山間地域の事業と事業規模の差が生まれていると考える。

その一方で，組合的な雰囲気の中で農家が主役となれるよう，株式会社いろどりは第三セクターを形態として選択した。同社は，商法法人の第三セクターとして営利追求企業活動に特化し，「彩」事業の一部を担当している。会社の資本金は上勝町が70％，残り30％はやはり第三セクター方式で町に設立された菌床シイタケを作っている株式会社上勝バイオが出資した。さらに，会社の運営費は，彩部会が5％，その他の全部の部会からの拠出金でまかなっている（横石，2007）。

「彩」事業では，農家，農協，株式会社いろどりの三者が所属の独立性を保ちつつ，一体となって運営している（図5）。株式会社いろどり，JA東とくしま上勝支所，農家が緩いつながりを形成している。これは，農家の「モチベーション」を鑑みた組織のデザインでもある。農家は自営業であり，行動の自己裁量が制限されるとモチベーションが制限される可能性が高いと考えるならば，「彩」事業で農家がおのおの事業家であり続けることがモチベーションに効果的であると考えられる。

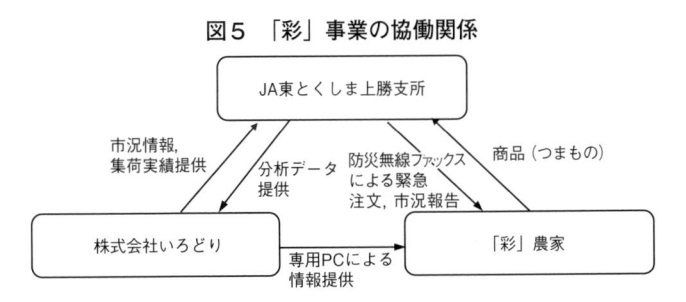

図5 「彩」事業の協働関係

出典：いろどり（資料を加工）。

4-2-2.「彩」の上勝町の社会面への効果

まず「彩」事業に参加する農家にとっての恩恵である。

第一に「彩」事業による収入を得られるようになった。年金暮らしだった高齢農家は「彩」で収入ができて、所得税を納めるようになった（横石, 2007）。

生きがいなど「ソフト」面での効果もみられる。特に「てご」と言って男性の手伝い作業の担い手であった女性にとっては、初めて自分でお金を稼いだ経験であることが多い。このことによって、自信と生きがいを持つようになった（横石, 2007）。また、「彩」事業を通じて料亭見学をしたり、外部の人とコミュニケーションの機会を持つことを通じて、「彩」事業に参加する農家は「垢抜け」（横石, 2007）してきた。筆者が参与観察で訪問した（2014 年 1 月 30 日）農家も大変積極的で、対人コミュニケーションにすぐれていた。

また、健康への効果は興味深い。上勝町は高齢化率が徳島県下一位であるにもかかわらず、2000 年には一人当たり医療費は 26 万 1,844 円と、徳島県の平均とほぼ同じ水準であり、県下で 48 自治体中 32 位であった（吉田, 2007）。「彩」事業参加農家に疾病は比較的少ないが、健康状態と疾病率の相関関係は確定が難しく、健康だから「彩」に参加しているのか、「彩」に参加していることで健康によい影響が出ているのかはっきりしない（稲葉, 2013）。「彩」事業参加農家は生活能力が高いが、疲労の自覚も高い（藤井・多田・岡久・松下, 2011）。

山口・近藤・柴田（2012）は、「彩」事業参加者ではない上勝町の高齢者においては有償労働と主観的幸福感に相関がみられなかったのに対し、「彩」事業従事者では正の相関がみられたことを報告し、「彩」事業参加農家の場合には、主観的幸福感を高める経済的要因以外の別の因子の存在を推察している。「彩」事業参加農家の場合には、「喜ばれることがある」の回答が多かった。

そして、「彩」事業の正の効果は上勝町全体に及んでいる。高齢者や女性達に仕事ができたことで、出番と役割ができ、元気になり、町の雰囲気が明るくなった。「葉っぱビジネス」の仕事が忙しくなってきたため、老人ホームの利用者数が減り、町営の老人ホームはなくなった（横石, 2007）。

大きな変化は、上勝町内の人間のつながりが変化したことである。悪口の対

図6 「彩」ブランド売上と上勝町人口の推移

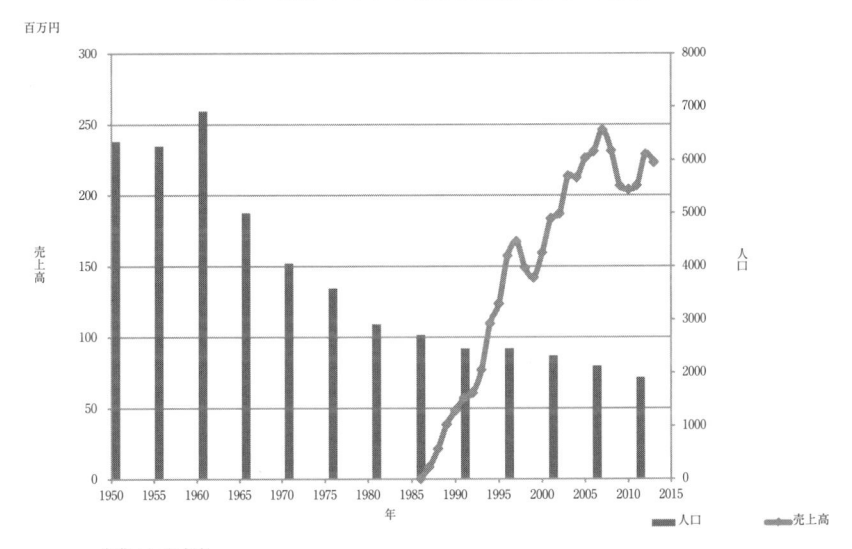

出典：いろどり。
注1：「彩」商品の販売はJA東とくしま上勝支所の担当であり，売上はJA東とくしま上勝支所のものである。
注2：図中の売上高は，「彩」商品の一部分である。

象だった嫁や近所の人たちは「彩」事業のパートナーとなり，一番大きく変わったのは，住民がみんな町のことを自分の問題として考えるようになったことである。「彩」のブランド力を上げるために，町の環境も自分が守らなければと考えるようになってきた（横石，2007）。

「彩」事業の成長にもかかわらず上勝町の人口減少が続いていることからもわかるように（図6），「彩」事業だけで上勝町が抱えるすべての問題を解決することは不可能である。長期的には，上勝町の人口を増加傾向にしていかなければ，町の持続性は困難になるが，それは「彩」事業だけで対応する規模の問題ではないということであろう。

しかし興味深いのは，上勝町の人口が減り続けているにもかかわらず，町に活気が出て，経済的，社会的効果が顕著にみられることである。2014年1月31日に上勝町「彩」を参与観察した際に，町役場で聞き取りを行ったところ，

「彩」事業参加農家の子供が上勝町に住んでいない場合でも，上勝町の親を頻繁にたずねて町の維持のための作業や「彩」事業に関わったりしていることが町の自治が維持されている一因であるとのことであった。

4-2-3. 「彩」事業のイノベーションプロセス

「彩」事業のイノベーションは，「つまもの」の市場を新規に作ったことである[8]。

「彩」事業が，上勝町内にとどまらず，全国に市場を持ったことによる経済効果は大きい。

「彩」は，技術的な新規性はあまりないが，社会的な側面では，料亭見習いがライフスタイルの変化から葉っぱを集める作業を嫌うようになった変化に見事に対応している。Röpke (2002) の「ニーズが顕在化している／財の供給がない」新規市場を作ったといえる。「彩」事業の着想までには，横石の学習期間が存在する。横石は上勝町農協（現JA東とくしま上勝支所）の営農指導員として，上勝町の農業振興を業務担当していた（横石, 2007）。横石は営農指導員として，上勝町の地理や集落の農業の内容を把握していた（横石, 2007）。例えば，農家一人ずつが最大限の力を発揮したら，どのくらいの生産量を見込めるかを，横石は経験から知っていた（横石, 2007）。また，他地域でも栽培できる一般の農作物では，上勝町の差別化が難しいことを経験から学んでいる。この経験からRöpkeの「能力」と「モチベーション」が作られたと考えてよいであろう。横石にとって，つまものを「単価が高い，毎日売上が上がり短期間で収入が得られ，季節に関係ない」（横石, 2007），持続性のある事業に育てる以外の，例えば一回性のイベントなどの選択肢は魅力的でなかったと思われる。「つまもの」は可能な選択肢の1つであった点でその選択は偶然であるが，常に差別化を考えていた横石が葉っぱに目を留めたのは偶然ではない。さまざまな外界の刺激のうちでも葉っぱに反応したのは，オートポイエーシス的なメカニズムとして見ることができよう。

また，農家との付き合いを通して，横石はモチベーションとして，上勝町農家が生きがいを得られることを切望するようになり，農家自身が主役となる事業を起こしたいと考えた。横石にとっては，高齢の農家が比較優位性を持てることが，事業内容の選択での与件であった。「つまもの」への着想は，この点

からもロジカルな偶然の選択である．上勝町農家は，1980年代初めから花木を始めており，農家の高齢女性たちは山の自然や葉っぱに関するノウハウを豊富に持っていた（横石，2007）．さらに，農家は花木を植える土地を所有しているという強みもある．

農業の知識を持っていることに加えて，横石の先見性と想像力は優れている．「彩」事業は，京都や大阪など，料亭文化が盛んな都市に近い立地条件を生かしている．また，料亭見習いのライフスタイルの変化に伴う，見習い作業に対する感情を感知した点で，先見能力があると考えられる．

横石と農家のコミュニケーションは密である．相互にポジティブに相手からの刺激を選択しているように見える[9]．横石にとって農家は事業のパートナーであり，共感を持つ存在である．横石は，「彩」事業参加農家の育成に質量ともに，平均をはるかに超える力を入れている．一例として，横石は「彩」事業参加農家に料亭見学をさせている．農家のモチベーションシステムを熟知していたため，料亭見学の効果も現実的に想像できていたのである．横石が上勝町で話していることを一流の料理人から直接その場で聞くと，農家への効果は遥かに高かった（横石，2007）．ネットワークシステムも，農家の負けず嫌いな傾向を的確に捉えたものである．

農家は，市場で値がつき売れ始めるという実績の積み重ねを持って，葉っぱも商品であると納得するようになった（横石，2007）．ルーマンのシステム論に拠ってみれば，農家側での自省によって外界からの刺激（横石が，つまものはビジネスであり，恥ずかしいことではないと農家に伝えること）が再吟味されて選択されたとみることができよう．

つまものの市場化を事業として成立させるために構築した販売方法，情報システムは，イノベーションプロセスの中で重要な役割を果たした．このシステムを使って市況を流し，かつ，注文をとるように活用した[10]．これによって市況に迅速に柔軟に対応できるようになり，商品としての価値を上げることになった．

供給過多で値崩れすることを避けるために，従来の農産物の取引と異なる，生産調整をよりよく行えるシステムを構築したことは，「彩」事業の中のイノ

ベーションである。つまもののユーザーは限定された数でしかないので、需要と供給のバランスや納期の管理は何を置いても重要だった。そこで、POS（販売時点管理）システムを取り入れた。商品にバーコードをつけて在庫数や出荷数を管理し、販売情報のデータを分析して市場の動向をつかむ。農家にもパソコンを設置して会社とイントラネットでつなぎ、販売情報を農家側にも伝えて商品の出荷調整を自ら行うようにした。

パソコンやタブレット端末で見る「上勝情報ネットワーク」で、農家はパソコンやタブレット端末を駆使し、全国の市場情報を分析して自らマーケティングを行い、栽培した葉っぱを全国に出荷する。「上勝情報ネットワーク」では自分が彩部会で何番目の売上を上げているかの順位が分かるようになっているなど、農家のやる気を出させる「ツボ」をついた情報を提供しており、モチベーションを刺激するようにしている。

5. 持続可能な地域経済への含意

地域活性型農商工連携のモデルとなっている「彩」事業は、利益追求の事業体が、地域の社会的課題（高齢者の生活の質）を事業内容の条件とし、彼らを事業のキーコンピテンシーとした点でユニークである。偶然と自発性に沿った発生プロセスのように見えるが、内部の自己準拠性に則ったプロセスであるとみることができよう。「彩」事業は、自治体の産業振興策と異なり、上勝町民全員に等しく効果が出るような事業デザインではない。しかし、この割り切りがあるからこそ、自治体とは異なったユニークな社会的効果を出せているとみられる。

（1）すでにハーシュマン（1961）は、農業の前方関連効果が薄く、連関効果を通じて刺激を起こす力がかけていると指摘している。
（2）また、シュンペーター（1993）が主張した5つの新結合のケース（1. 新しい財の生成、2. 新しい生産方法の導入、3. 新しい販路の開拓、4. 原料あるいは半製品の新しい供給源の獲得、5. 新しい組織の実現）から考えても、価値の書き換えの段階で組織の変更、販路の開拓など、なんらかの「新結合」が伴うことは容易に想定されるであろう。この点

について，Röpke（2005）は，Kim and Mauborgne の「市場の再構築」は生産手段の新結合にあたると指摘している。
（3）例えば，地域外から産業を誘致する方法をとる場合には，当てはまらないであろう。
（4）「企業家」とも表記される。本稿では「起業家」の表記を用いる。
（5）たとえば，生物システムと社会システムは多くの点で類似しているように見えるが，生物システムをそのまま社会システムに応用することには問題がある。
（6）事業を指す場合は「彩」，会社は「いろどり」と表記する。
（7）現在，卸売市場の取引は予約型が主流となっており（細川, 1993），「彩」はこの流れに乗ったといえる。
（8）「彩」事業の前にも，料理に添える花物は市場に出されていたが，数量が少なく，また，季節限定販売であり（横石, 2007），つまもの市場が存在していたとはいえない。
（9）2014年1月30日，31日の参与観察中にも観察された。
（10）防災無線ファックスは停電の時でも動き，役場のシステムなので通信にもコストがかからない。

<参考文献>

石川和男（2013）「持続可能な地域社会創造の取り組み：徳島県勝浦郡上勝町における「彩」事業を中心として」,『専修大学社会科学研究所月報』, No. 601 & 602, pp. 128-143。

石川公彦（2013）「経済社会を創造する『まちづくりの論理』」,『一橋ビジネスレビュー』, Vol. 61, No. 2, pp. 56-71。

稲葉陽二（2013）「高齢者の社会参加で医療費低減：徳島県上勝町のケース」,『保険師ジャーナル』, Vol. 69, No. 6, pp. 462-466。

入谷貴夫（2008）『第三セクター改革と自治体財政再建』, 東京：自治体研究社。

いろどり（n.d.）葉っぱのまち上勝町いろどり：いろどりストーリー。Avaiable at http://www.irodori.co.jp/asp/nwsitem.asp?nw_id=2&design_mode=0, Accessed April 29th 2015。

笠松和市・佐藤由美（2008）『持続可能なまちは小さく，美しい：上勝町の挑戦』, 京都：学芸出版社。

経済産業省（2004）『通商白書2004：「新たな価値創造経済」へ向けて』, 東京：ぎょうせい。

斎藤修（2012）『地域再生とフードシステム―6次産業，直売所，チェーン構築による革新―』, 東京：農林統計出版。

櫻井清一（2010）「農・工・商・官・学の連携プロセスをめぐる諸問題」,『フードシステム研究』, Vol. 17, No. 1, pp. 21-26。

関満博（2014）『6次産業化と中山間地域：日本の未来を先取る高知地域産業の挑戦』, 東京：新評論。

武石彰・青島矢一・軽部大（2012）『イノベーションの理由：資源動員の創造的正当化』, 東

京：有斐閣。
谷本寛治（1993）『企業社会システム論』，東京：千倉書房。
――（2009）「ソーシャル・ビジネスとソーシャル・イノベーション」，『一橋ビジネスレビュー』，Vol. 57, No. 1, pp. 26-41。
中村良平（2014）『まちづくりの構造改革：地域経済構造をデザインする』，東京：日本加除出版。
西口敏宏（2009）『ネットワーク思考のすすめ：ネットセントリック時代の組織戦略』，東京：東洋経済新報社。
藤井智恵子・多田敏子・岡久玲子・松下恭子（2011）「山間地域で主体的に運営する産業に従事している高齢者の保健行動」，『JNI』，Vol. 9, No. 2, pp. 15-24。
細川允史（1993）『変貌する青果物卸売市場：現代卸売市場体系論』，東京：筑波書房。
堀島勇夫・望月正光（2007）『第三セクター：再生への指針』，東京：東洋経済新報社。
山口静枝・近藤昊・柴田博（2012）「農村地域の自立高齢者における productive activities が主観的幸福感に及ぼす影響」，『応用老年学』，Vol. 6, No. 1, pp. 59-69。
横石知二（2007）『そうだ，葉っぱを売ろう！過疎の町，どん底からの再生』，東京：ソフトバンククリエイティブ。
吉田俊幸（2007）「高齢者農業の可能性とその社会的意義―中高年層での新規就農，就農の強まり―」，『地域政策研究』，Vol. 9, No. 2 & 3, pp. 17-33。
Bourdieu, P. and Wacquant, L. J. D. (1992) *An invitation to reflexive sociology*, Chicago : University of Chicago Press.
Christensen, C. M. (2000) *The innovator's dilemma: The Revolutionary National Bestseller That Changed The Way We Do Business*. New York: HarperBusiness.
Combé, N. (2008) *Der Knowing-Doing-Gap im Innovationsprozess postindustrieller Gesellschaften, Marburg*: Mafex（BOD）.
Feldhoff, T. (2013) 'Shrinking communities in Japan: Community ownership of assets as a development potential for rural Japan?', *Urban Design International*, Vol. 18, No. 1, pp. 99-109.
Haga, K. (2013) *Innovations- und Evolutionsdynamik in demographisch alternden Gesellschaften*, Marburg: Metropolis.
Hamel, G. and Prahalad, C. K. (1994) 'Competing for the Future', *Harvard Business Review*, July-August, pp. 122-128.
Hirschman, A. O. (1958) *The strategy of economic development*, New Haven: Yale University Press.（麻田四郎訳『経済発展の戦略』巌松堂出版，1961 年）
Karlsson, C. (2012) 'Entrepreneurship, Social Capital, Governance and Regional Economic Development', *SIR Electronic Working Paper No. 2012/6*. Available at https://www.6th.sel/mam/forskning.nsf/attachments/WP6_pdf/$file/WP6.pdf, Accessed July

22nd 2014.
Kim, W. C. and Mauborgne, R. (2005) *Blue Ocean Strategy*. Boston, MA: Harvard Business School Press.
Krugman, P. R. (1991) *Geography and Trade*. Cambridge, MA: MIT Press.
Luhmann, N. (1984) *Soziale Systeme: Grundriss einer Allgemeinen Theorie*, Frankfurt am Main: Suhrkamp.
―― (2002) *Einführung in die Systemtheorie*, Heidelberg: Carl-Auer-Systeme Verlag.（土方透監訳『システム理論入門―ニクラス・ルーマン講義録[1]』新泉社，2007年）
Röpke, J. (2002) *Der lernende Unternehmer*, Norderstadt: BOD.
―― (2005) Value innovation. How to create uncontested market space and make the competition irrelevant, Available at http://etc.online.uni-marburg.de/etc1/008.pdf, Accessed April 25th 2015.
――and Xia, Y. (2007) *Reisen in die Zukunft kapitalistischer Systeme*, Norden stadt: BOD.
Schumpeter, J. A. (1993) *Theorie der wirtschaftlichen Entwicklung*, (8. Aufl.), Berlin: Duncker & Humblot.（塩野谷雄一，中山伊知郎，東畑精一訳『経済発展の理論（上下）』岩波書店，1977年）
―― (2010) *Konjunkturzyklen: Eine theoretische, historische und statistische Analyse des kapitalistischen Prozesses*, Göttingen: Vandenhoeck & Ruprecht.
Smallbone, D. and North, D. (1999) Innovation and new technology in rural small and medium-sized enterprises: some policy issues, *Environment and Planning C*, Vol. 17, No. 5, pp. 549-566.
Westley, F., Zimmerman, B. and Patton, M. (2006) *Getting to Maybe: How the World Is Changed*. Tronto: Random House of Canada.（東出顕子訳『誰が世界を変えるのか：ソーシャルイノベーションはここから始まる』英治出版，2008年）

3 韓国の社会的企業の'制度化を通じた'育成に関する考察
――「社会的企業育成法」を巡る現状と課題を中心に

金　仁仙

対外経済貿易大学公共管理学院（常勤）講師

【要旨】
本研究は韓国の社会的企業の発展と現状について踏まえた上で，更に政府支援の実態について考察し，それを巡る挑戦と機会，戦略と課題について考察する。韓国の「社会的企業育成法」に基づいた政府主導の社会的企業の育成は，短期間にわたる社会的企業数の飛躍的な増加によって脆弱階層に職業と社会サービスを提供するなど，社会問題の解決に大きな貢献を果たしたが，同時に，社会的企業の自立性を制限することによって，社会的企業の生み出す社会価値の限定と政府依存度の向上，および経営体質の悪化による持続可能性の低下という問題をも生み出した。これらを踏まえた上で，本研究の結論においては，今後の社会的企業の持続的な発展を向け，法律または制度，金融市場，そして中間支援組織の側面から期待される具体的な取り組みについて提示する。

キーワード：社会的企業，韓国，社会的企業育成法，社会的経済

1. 社会的企業の台頭

　全世界における経済危機と財政危機，それに伴う失業と貧困の問題を解決する新たな経済システムとして社会的経済（Social Economic）が注目されると，社会的企業は社会的経済を構成する1つの領域として，産業社会における社会問題を解決する過程で発生した市場および政府機能の失敗を修正・補完するために取り上げられたものである（이주호, 2013）。社会的企業は，70年代以降の

欧州を中心とする資本主義市場経済体制の限界において誕生し，90年代以降の欧米国家の福祉国家の危機と共にその重要性が強調され，2000年代に入ると，世界的に社会的企業や社会的企業家（social entrepreneur）という概念が，政策的・実践的・学術的にも認知されるようになった。社会的企業の台頭の背景には，非営利組織の収益事業の比重の高まり，市場や政府の供給しにくい公共サービス分野の提供者としての役割と，社会的排除から社会的包摂へという社会的なつながりの回復に対する役割の増大，企業の社会的責任に対する関心と役割の向上等が挙げられる（塚本, 2012）。

　社会的企業に関する財界，政界と学界の関心が高まる中，それに関する定義は国によって多少の違いが存在する。それは，社会的企業が社会的課題の解決を主な目的として事業を行う上で，各社会における問題認識は，社会的企業の活動する社会の歴史と文化，経済および政策などによって異なりうるからである。イギリスの内閣府は，社会的企業について，「組織の主な目的は社会的なものであり，株主およびオーナーの利益を最大化するために経営を行うのではなく，利益を事業および地域に再投資する企業」と説明し（조영복, 곽선화 옮김, 2011），ヨーロッパにおける社会的企業研究者のネットワークである EMES（the Emergent of Social Enterprise in Europe）は，社会的企業を「地域社会への利益という明白な目的に直接的に関係付けられる財やサービスを提供する，私的な利益を目的としない（not-for-profit）組織である。さまざまな立場の利益関係者を意思決定に巻き込むという集合的活動に依拠し，自律性に高い価値を置き，社会的企業の行動と関連付けられる経済的リスクを負う」ものと定義する（곽관훈, 2012）。また，OECD（1999）は，社会的企業について「企業的な方式によって組織されるが，企業の追求する主な目的が，利益の極大化ではなく，経済的かつ社会的なものであり，社会における除外と失業の問題に対し革新的な解決策を提示することのできる，公益活動を展開する民間組織または企業」と定義を行い，社会的企業の顕著な特徴について，失業や社会的排除の問題に対し革新的でダイナミックな解決策を見出すことや，社会的紐帯（social cohesion）を強め持続可能な経済開発に貢献することと説明する。以上の内容を踏まえると，社会的企業は，一般的に，「社会的・経済的目的を同時に持ち，

経済的手段を用いて社会的目的を実現する新たな経済主体」と理解することができる。なお，社会的企業が社会経済に齎す主な役割については，1脆弱層の雇用を含め，経済活動を通じた経済的価値の創出，2多様化かつ複雑化した社会的要求の満足，3地域社会の再建，4良質の公共サービスの提供と福祉社会の実現，5市場倫理と社会統合の促進などと集約され（김인선&노희진, 2013），従って，社会的企業は，政府の社会と経済に関する政策の実現に貢献し，特に新たな福祉社会のモデルを実現する上で重要な役割を果たすことと理解出来よう。

　一方，国内における社会的企業に関する研究は，その対象と内容において，従来の欧米国家を対象に，社会的企業の概念と役割に分析の焦点を当て，社会的企業の育成の必要性を強調してきたことから，近年においては，隣国である韓国の社会的企業の発展について，その現場における実態を視野に入れ，実証的な分析を行う研究も増えてきた（駒崎・小倉, 2013）。その背景には，日本と共に，2000年に入ってから社会的企業の概念を本格的に導入し始めた韓国においては，2007年にアジア初で「社会的企業育成法」を制定することによって飛躍的な発展を果たし，2014年10月には「第7回・社会的企業世界フォーラム（Social Enterprise World Forum）」がアジアでは初めて韓国ソウルで開催され，国際社会において「社会的企業育成法」制定その後の実績が評価されるなど，国際社会からも多く注目されることがある。桔川純子（2010）と柏井（2013）は，韓国の社会的企業育成法の制定を市民社会の新たな展開として捉え，それが現代社会に齎す意義について史的観点から明らかにした。柏井（2013）は，韓国における社会的企業の概念の萌芽について，1910年以降の日本帝国主義の朝鮮侵略以降にたどり，結社体を問い共同社会経済への着実な積み重ねの連続とみなし，「社会的企業育成法」の立法とその展開は，市民運動家の闘いによって得られた結晶物として捉える（桔川, 2010）。白井（2008）と橋本（2011），加藤（2013）などは，韓国の「社会的企業育成法」の立法における主な背景に着目し，つまり，韓国では，1990年代末の経済危機とその後の労働市場の柔軟化の中で，雇用情勢が悪化し，その問題を解決すべく，国の雇用創出・拡大政策の一環として社会的企業育成政策が展開され，社会的企業が雇用

創出政策の主な担い手となったことに焦点を当て議論を展開する。橋本（2011）は，韓国における労働総合型社会的企業は，ワークフェアや積極的労働市場政策が遂行される上での具体的な事業の担い手として位置づけられ，韓国における社会的企業は政府の思惑と市民運度の要求がせめぎ合いながら制度化が進められ，その背景には，韓国の市民運動の厚みがあることがその一因であると示し，この点では，桔川（2010）と柏井（2013）とも観点を共にする。

　韓国の法律の制定による社会的企業の育成といった制度化による社会経済の実現が持つ意義にもかかわらず，韓国の社会的企業の短期における飛躍的な成長は，それが民間，または市民社会の主導というよりは，政府主導の制度化によって持たされたことであるとの特徴から問題も少なくない。それらは，政府と与野党共通の核心事業である「社会的企業」育成事業が量的膨張ばかりか，社会的価値追求と本来の目的を喪失し，社会サービス提供，地域社会貢献に限界があること，なお，事業基盤が弱い上，助成金への依存傾向が強いことに集約される（박태정, 2011；이희수, 2011；곽관훈, 2012；김상희, 2009）。韓国国会立法調査処による現場報告書は，「わが国の社会的企業の形態は雇用創出類型を中心に運営されている」とし「このような雇用事業中心の社会的企業育成政策では，社会的脆弱階層に対する社会サービス提供という本来の社会的価値が過小評価されている」，また，「雇用創出事業中心の量的拡大は，現在汎政府次元で施行されている『財政支援雇用事業』と大差ないばかりか，拡大する社会的企業もほとんどが小規模零細業者の可能性が高く，さらに政府支援に依存するという悪循環になりかねない」と指摘する（손을춘, 2014）。이정봉（2010）は，韓国において社会的企業は，社会サービスの職場が産業化へと進む政策的手段として捉えられ，社会的企業育成法によってその支援を保障し，社会的企業における労働条件において，企業内の脆弱階層の労働者の比率が高く（58％），社会的企業労働者たちの賃金水準が低く，また，雇用不安定要素が高い（非正規職員が約73％で，女性労働者10人の中8人の78.9％が非正規職員である）との問題を招いたと指摘する（이정봉, 2010；姜と落合, 2011）。駒崎と小倉（2013）は，韓国の社会的企業に対するヒアリング調査から，韓国における社会的企業の構造的問題点として，韓国における社会的企業の活動が国の政策の一環として展

開されていることから，法律に基づいて支えられているが，法定上の諸条件を満たせない団体は社会的企業としての認定を受けられず，不安定な環境の中で活動をせざるを得ず，更に，雇用創出・拡大政策の主体として，雇用の質が「短期間・低賃金・非熟練」にならざるを得ない状況であると指摘し，今後の発展過程において，国の支援のあり方や中間支援団体のあり方を再検討し，問題を解決してゆく必要について訴える。

　以上から，韓国の政府主導の制度化による社会的企業の育成は，短期間における社会的経済を担う主体の増大と社会一般における認識の拡散に影響する一方で，社会的企業に関する十分な考察に基づいた社会的合議がなされず，政策の潮流の下で制度的推進によって展開は結局"制度的同型化（institutional isomorphism）"の問題を招き（장원봉, 2008；김성기, 2009），更に，実務部署の業務量の増大と社会的企業の多様化を妨げる要因として指摘される（김혜원, 2009）。この点を踏まえた上で，本研究では，韓国の社会的企業の発展について，「社会的企業育成法」を巡り，その実態を把握し，それに基づき，今後の行き先について考察することを目的とする。そのために，まずは，韓国の社会的企業育成法の立法を巡り，韓国における社会的企業の勃興の背景について考察し，その過程における問題の原点に迫る。その上，韓国における社会的企業の概念，それに基づいた実態，そして，それを支える支援体制について把握し，韓国の社会的企業の発展を巡る全体像をつかむ。次に，韓国の社会的企業の発展に関しては，上述したように，関連制度が持つ構造的な限界から社会的企業の持続的な発展のためのいくつかの重要な議論されることから，それらに基づく韓国の社会的企業の発展を巡る挑戦と機会について述べ，そのため，現時点における政府の新たな取り組みや，それによって期待される効果について把握する。最後に，以上の韓国の社会的企業の育成を巡る経験から，今後の社会的企業の持続的発展のための戦略と課題について述べる。

2. 韓国社会的企業の発展背景

　韓国において社会的企業は，アジア通貨危機以降，政府によって社会の格差

(韓国語では両極化)と職業,および社会福祉の問題を解決する重要な手段として用いられたものとして理解できる。1997年にアジア通貨危機の影響を受け,韓国経済の成長率は98年には前年より10ポイント下落し-5.8%まで下落し,1万を超えていた一人当たりGDPは6,823ドルまで落ち込んだ。その影響を受け,失業率は97年2.62%から翌年には6.95%の7%まで激増すると,政府が緊急対策で展開した労働柔軟化政策は,非正規,低賃金労働者のみを対象にしていたため,女性と青年など,いわば,不安定な勤労者とワーキングプアを増大させ,労働市場の両極化という問題を招いた。韓国統計局によると,正規社員の比率は,アジア通貨危機直前である1996年の53.6%から2000年には47.7%まで減少,非正規社員の比率は,同期において43.2%から52.3%まで増加し,社会格差を表すジニ係数は,1996年0.29から2007年には0.35にまで急増した。政府主導の構造改革の結果,1999年にはGNP上昇率の10%を達成し,いわば,経済のV字回復を記録したが,当時の脱産業化に伴う高学歴の未就職者(韓国語ではペクス)と働く貧困層の増加は,労働市場の両極化の更なる深化と韓国社会を雇用無き成長時代に導いたのである。社会格差を表す指標として用いられる全国全世帯五分位所得倍率(所得水準上位20%の所得を所得水準下位20%に割った比率)は,2003年の7.14から2007年には7.27まで拡大し,雇用無き成長の現象を表す雇用係数(実質GNP10億ウォン当り就職者数)は,アジア通貨危機直前の40台から,99年には38に,2004年には32まで下落した(桔川,2010)。社会的・経済的な不安の拡大に加え,離婚率の急増と高齢化,少子女化の進展は,社会サービスに対する需要の拡大と多様化を齎したため,これも,政府の'脆弱階層に対する職業の提供を通じた社会サービスの供給',いわば,社会的企業の育成を後押しする重要な背景となった。

　従って,韓国において社会的企業は,経済不況から回復した後にも依然として改善されていない就職問題を社会・経済の持続的発展を妨げる構造的問題として捉え,その根本的な改善に向け制度化を進める過程で取り入れられたものと理解できる(桔川,2010;조영복,2011;이대영&이상희,2009)。政府は,アジア通貨危機直後の急増した失業率に対する緊急対策として取り入れた「公共勤労事業」をはじめ,2000年には,1999年に立法をこぎつけた「国民基礎生活

保障法（国基法）」に基づき，勤労能力を持つが，所得と財産が基準以下であり，扶養義務者も経済的な能力を持っていない人に対し，政府（保険福祉部）が指定する「自活後身機関」の自活事業への参加を条件とし自活給与を提供する「自活支援事業」を展開した（国基法施行令第8条）（友岡有希, 2015）。2003年には，ノ・ムヒョン政権下の労働部（現在の雇用労働部）がおよそ73億ウォンの予算を投入し，脆弱階層に対する雇用の創出と社会サービスを提供する「社会的職労事業」を実施した。また，2000年代に入り，失業問題が顕在化する中，高学歴青年間の学歴競争が展開されたため，政府は高齢化や青年労働力の純減等の人口問題を国家危機と捉えて，技術教育改革と社会貢献・起業支援を通じた雇用の柔軟化・差別化を推進することで，さらに国際競争に立ち向かう意向を含めた「青年雇用促進特別法（2004～13年）」を打ち上げた（金早雪, 2011）。しかし，政府にて推進されたこれらの措置は，政府依存度が高く，その結果，雇用を創出するものの短期，低賃金の職業がほとんどを占め，また，社会サービスを提供する対象が脆弱階層に限定され，政策の支援を得るべく財政基盤を支える中産階級までサービスを拡大しようとすると財源がないとの問題を抱えていた。そのため，持続性，安定性の側面から，問題の根本的な改善を必要とする認識が政府と民間において拡大していく中，2005年にはハンナラ党議員が発議した社会的企業に関する法案が国会に提出されることによって，社会的企業の法制化を巡る議論が国会と労働部の間で本格的に展開された。そして，2006年12月には「社会的企業を支援し，我が社会で十分に供給されていない社会サービスを拡充し新しい就労を創出することにより，社会統合と国民の生活の質の向上に寄与する」ことを目的とする（第1条）「社会的企業育成法」（法律第8217号）が制定され，翌年7月から実施されるようになった。これまで政府による‘公共勤労事業’，‘自活支援事業’，‘社会的職労事業’など，脆弱階層の雇用を創出することに焦点が当てられた事業は，「社会的企業育成法」といった具体的な法律に基づき，‘社会的企業’という新たな主体によって推進されるようになったのである（図1）。

図1 アジア通貨危機以降の主な社会現象と政府の失業対策

年代	1990年代後半	2000年代前半	2000年代後半
社会現象	失業率の急増2.62％（1997年）→6.95％（1998年）労働市場の二極化（正規・非正規），中壮年世代の失業	社会サービスに対する需要の増大と複雑化 政府の福祉抑圧強化	
失業対策	1998年 公共勤労事業	2000年 自活支援事業	2003年 社会的雇用創出事業 2004～2013年 青年雇用促進特別法	2007年 社会的企業育成法制定・実施

出所：桔川純子（2010），조영복（2011），이대영&이상희（2009）などを参考に筆者作成。

3. 韓国社会的企業の概念と現状

「社会的企業育成法」の制定から5年目を迎える2011年12月には，社会的企業の更なる活性化を目指し，その一部を修正した「社会的企業育成法の一部改正案」が国会からの可決を経て，2012年2月1日から施行されるようになった。本節では，2011年に修正された「社会的企業育成法」および「社会的企業育成法執行令」を基に，韓国の社会的企業の概念と現状を把握する。

同法は，社会的企業を脆弱階層[1]に社会サービスまたは仕事場（就労）を提供し，地域社会に貢献して地域住民の生活・命の質を高めるなどの社会的目的を追求しながら，財貨およびサービスの生産・販売など営業活動をする企業として第7条によって認証を受けた者（第2条）と定義し，認証を受けるためには，その第8条で定める，①「民法」による法人・組合，「商法」による会社・合資組合，「特別法」により設立された法人または非営利民間団体等の組織形態をとること，②有給の労働者を雇用し営業活動をすること，③脆弱階層に社会サービスまたは仕事場（就労）を提供し地域社会に貢献して地域住民の生活・命の質を高めるなどの社会的目的の実現を組織の主たる目的とすること，④利害関係者が参加する意思決定構造を整えること，⑤営業活動を通じて一定の収益を出していること，⑥定款や規約などを定めていること，⑦「商法」による会社・合資組合の場合，会計年度別で配分可能な利潤が発生した際には，

利潤の3分の2以上を社会的目的のために使用することなどの要件を満たさなければならない。また、以上の認証条件を一部満たさないが、社会的企業を志す事業者を予備社会的企業（最大2年限度）として認める。2007年7月に50社あった社会的企業は、2014年までに30回の雇用労働部長官の認証審査を経て、1,251社に増加（認証率約50％弱、予備社会的企業1,466社）し、25％の社会的企業がソウル、京機道、仁川などの首都圏に集中している。

以下の表1は、韓国の社会的企業の実態を、「社会的企業育成法」と「施行令」で定める組織形態、社会的目的性、活動分野別に表したものである。韓国の社会的企業は、商法上の会社を除くほとんどが非営利性の組織形態をとっており、雇用創出を目的とする社会的企業が約7割程度を占め、混合型までを含むと、その割合は全体の約8割を超える。また、社会的企業が活動する分野に関しては、製造業を含むその他の分野を除き、環境と文化関連の分野において活動する社会的企業が高い割合を占める（表1）。

社会的企業の財務と事業費の構造、および売上と利益の分布状況からは、社会的企業の経済的成果について把握することが出来る。곽선화（2011）が491社の社会的企業を対象に調べた結果、2010年末を基準に、資産規模が10億

表1　2014年度韓国社会的企業の組織形態別，社会的目的別，活動分野別分布

（単位：％）

組織形態別	比率	社会的目的別	比率	活動分野別	比率
商法上の会社	54.1	雇用創出型	69.3	社会福祉	8.1
民法上の法人	19.3	社会サービス型	4.9	環境	15.6
非営利民間団体	9.0	地域社会貢献型	1.8	保健	1.0
社会福祉法人	8.0	混合型	12.4	文化芸術	15.5
営農(漁)業組合法人	4.0	その他	11.7	山林保全	0.2
農(漁)業会社法人	2.3	―	―	家事・看病	6.4
共同組合	0.9	―	―	保育	1.4
社会的共同組合	1.2	―	―	教育	7.0
その他の法人・団体	1.2	―	―	その他	44.8

注：「社会的企業育成法執行令」に基づき、雇用創出型とは、全体雇用者の中、脆弱階層の雇用比率が50％以上の社会的企業を、サービス提供型はサービスの受ける者の中、脆弱階層の比率が50％である社会的企業を、混合型は上述した2つの目的を追求しながら、脆弱階層の雇用比率とサービスの受ける者の中、脆弱階層の比率がそれぞれ30％である社会的企業のことを指す。
原典：韓国社会的企業振興院（2014）

ウォン以上の社会的企業は全体の12.6%に過ぎず,それに対し,資産規模の5,000万ウォン未満の社会的企業は全体の31.0%を占める。資本金の場合,1億未満の社会的企業が全体の半分程度(43.6%)を占める一方,3億ウォン以上の社会的企業が占める割合は16.7%に過ぎず,更に17.3%の社会的企業は資本蚕食状態であることが分かる。また,100億ウォン以上の負債を抱えた社会的企業が25.3%も存在し,負債の無いと応えた企業は7.3%に過ぎない。同調査に基づき,社会的企業の売上および利益の分布を見ると,売上げの場合,30億ウォン以上の社会的企業は4.5%と少ないことに対し,1億未満の場合が19.6%を占め,5社の社会的企業のうちの1社の社会的企業は1億未満の売上げを記録していることが推測できる。営業利益の場合,黒字を記録した社会的企業は全体の16.1%に過ぎず,5つのうち1つの社会的企業が赤字を記録しており,1億ウォン以上の赤字を記録した社会的企業は69.5%にも及ぶ。純利益においては,1億ウォン以上の黒字を実現した社会的企業が全体の11.9%を占めるが,営業利益において黒字を実現していないにもかかわらず,純利益において黒字を実現することができたのは,政府および大企業による財政支援が,営業利益以外の純利益に影響したからであり,このような財政支援にもかかわらず,19.9%に至る社会的企業は,依然として純利益の赤字を免じることがで

表2 2010年末の韓国社会的企業の財政状況

(単位:ウォン)

売上	比率(%)	営業利益	比率(%)	純利益	比率(%)
5千万未満	10.2	−5億以上	9.2	−5千万以上	2.5
5千万〜1億	9.4	−5〜−3億	12.1	−5千万〜1億	8.0
1億〜2億	15.5	−3〜−2億	11.3	−1千万〜0	9.4
2億〜3億	15.1	−2〜−1億	24.0	0〜3千万	13.3
3億〜5億	14.3	−1〜−0.5億	12.9	3千万〜5千万	31.8
5億〜10億	16.5	−0.5〜0億	14.3	5千万〜1億	15.8
10億〜20億	10.8	0〜0.5億	10.0	1億〜3億	10.7
20億〜30億	3.7	0.5〜1億	2.0	3億〜5億	8.0
30億〜50億	3.3	1億以上	4.1	5億〜10億	0.4
50億以上	1.4	—	—	—	—

原典:곽선화(2011)

きなかった（表2）。そのため，「社会的企業育成法」では，社会的企業が事業活動を通じ新たに創出した利益を社会的企業の維持拡大に再投資することを示しているが，韓国の社会的企業の中では，営業利益を実現した企業がほとんどなく，実際に社会的目的のため再投資の義務に該当する企業はほとんどないと理解できる。

　社会的企業の社会的成果に関しては，雇用水準と社会サービスの提供の規模，そして利益の再分配の水準から把握することが出来る（雇用労働部，2010）。社会的企業による雇用従業員者数は，2007年の2,539名から2014年12月には27,923名まで増加し（機関平均雇用数は22.3名），その中，脆弱階層の雇用者数は，1,403名から15,815名にまで増え（全体雇用の57％），また，2014年の脆弱階層の社会サービス受け率は70％を超える（韓国社会的企業振興院，2014）。また，韓国国会立法調査処による調査（2014）によると，韓国の社会的企業の中では，従業員数の30人未満の組織が全体の83.1％にも達し，これらの平均従業員数は12.7人程度でほとんどが零細業者と言える（손을춘，2014）。従って，以上からは，韓国の社会的企業の場合，その規模が小さく，政府への依存度が高い上，財政状況が悪化していないことが分かる。

4. 韓国社会的企業の支援体制

　社会的企業の発展のための政府支援策について把握することは，国別の社会的企業の育成モデルを理解する上で重要である。本項では，韓国政府の社会的企業の育成のための支援策について，主に「社会的企業育成法」に基づき，制度，財政，事業および運営，官民協力の側面から把握する（内閣府，2011；조영복，2011）。

4-1. 制度的支援

　韓国は社会的企業に対する支援の選択と集中を通じ，その効率を高めるための制度的基盤として，社会的企業認証制度を採択する。社会的企業を運営しようとする者は，雇用政策審議会の審議を経て，雇用労働部長官によって認証を

受けなければならない（第7条）。そして，認証を受けた社会的企業は，同法の第10～16条に基づき，①社会的企業の設立および運営に必要な経営支援，②専門人材の育成のための教育訓練，③施設費の融資，④優先的購買と租税減免，および社会保険料の補助，⑤社会サービスの提供における財政等の支援，⑥連携企業[2]に対する租税減免などの支援を受けることができる。予備社会的企業に対しては，人件費，専門人材用人件費，経営コンサルティング，事業開発費の分野に限り支援を行う。なお，認証を受けた社会的企業は毎年事業報告書を提出することが義務付けられ（第17条），一定の基準にそぐわない場合には認証が取り消され（第18条），社会的企業ではない者は社会的企業または類似する名称の使用を禁ずる（第19条）。

4-2. 財政支援

　韓国において社会的企業に対する政府の財政支援は，直接的方式を中心に，人件費の補助と社会保険料の支援，経営コンサルティング費支援と事業開発費の支援，および税制支援などが展開され，その中でも新規採用に対する人件費の補助が重要な内容となる[3]。「社会的企業育成法」に基づき，脆弱階層の雇用に対しては，3年間（予備社会的企業の場合2年）で月932,000ウォンを支援し，会計，マーケティングなどの専門人材の採用においては，企業当たり3名に限り最大3年間，月150万ウォン限度の人件費を支援する。2010年からは，人件費補助事業に参加していない社会的企業に対し，最大4年間の事業者負担分の社会保険料を支援している（労働者全員の保険加入および政府が規定する最低賃金以上の賃金を支払うことの条件付き）。経営支援の側面からは，労務や会計に関する専門的コンサルタントを受けるための経費を，最大3年間，2,000万ウォンを支援する。また，2010年からは，地方自治体と連携し，1年間で最大7,000万ウォンの事業開発費を支援している。税制支援に関しては，認証から4年間，法人税と所得税の50％の減免を実施しており，さらに連携企業の社会的企業を支援するための支出について，指定寄付金と見なし，法人所得の5％範囲内で全額損金算入を認めている。一方，社会的企業に対する間接的支援としては，政府による優先購買程度が挙げられる。

4-3. 事業支援

　社会的企業に対する支援事業が多様化，複雑化していくと，その体制的かつ効率的な推進を目指し，2011年2月には社会的企業に対する支援事業を専門的に行う公共機関である'韓国社会的企業振興院'(以下'振興院')が「社会的企業育成法」第20条に基づき，社会的企業と行政の連絡調整を果たす中間支援組織として，①社会的企業(家)の養成と社会的企業モデルの発掘および事業化の支援，②社会的企業に対するモニタリングおよび評価，③業種，地域および全国単位ネットワークの構築と運営，④社会的企業の統合ホームページおよびシステムの構築と運営，⑤経営・技術・税務・労務・会計などの改善のためのコンサルティング支援，⑥国際交流，⑦法律または法令によって委託された社会的企業に関する事業などを展開する。それに加え，'振興院'は，社会的企業の設立および運営に必要な専門情報の提供と人材育成，労働者の能力向上を支援する事業を展開する[4]。

4-4. 官民協力支援

　近年，韓国政府は，社会的企業の活性化に向け，社会的企業の政府機関および地方自治体との連携を更に強めている。政府は公共機関および政府部署において社会的企業を活用した事業展開を促進している。例えば，教育部は放課後の塾関連事業を展開している社会的企業を支援しており，文化体育観光部は雇用労働部と協力し，文化分野で活動する社会的企業の育成を支援する(이희수, 2011)。更に，予備社会的企業の育成を向け，行政自治部は，地域共同体働き口事業を，産業通商資源部は，コミュニティビジネスモデル事業を，農林畜産食品部は農漁村共同体会社育成などの活動を取り組んでいる。

　一方，「社会的企業育成法」においては，地方自治体において市・道別の社会的企業支援計画を策定し施行することを義務付け(第5条の2)，社会的企業の全国的拡散と地域経済への貢献を促進している。2009年からは，自治体の社会的企業に対する優先購買の実績を，行政自治部の自治体に対する助成金額を決定する判断資料として扱うことにしている(박찬임など, 2009)。

図2 韓国の社会的企業支援体制

支　援	内　　容
制度的支援	（予備）社会的企業認証制度：一定の条件を満たす企業は政府からの認証を獲得し，社会的企業として関連支援を受けることが出来る。
財政支援	―直接的支援：人件費，保険料，経営コンサルティングおよび事業開発費 ―間接的支援：社会的企業および連携企業に対する税制減免
事業支援	社会的企業振興院の設立（2010）：社会的企業家および人材の育成，事業化などの支援
官民協力支援	―部署別事業展開時の社会的企業の活用 ―自治体の評価項目に社会的企業に対する優先購買を含む

出所：内閣府（2011）と韓国社会的企業振興院ホームページを参考に筆者作成。

5. 韓国の社会的企業の育成における挑戦と機会

　本節では，以上で説明した「社会的企業育成法」の主な内容と構造的特徴に基づき，それに関する主な議論を中心に，今後の韓国社会的企業の展開における挑戦と機会について考察すべく，韓国国内における最新情報を用いて政府の新たな取り組みなどについて把握し，今後，韓国社会的企業の行き先をつかむ。

5-1. 中央政府による社会的企業支援政策の見直し

　2000年代後半，政府と与野党共通の核心事業である社会的企業の育成事業が，量的膨張ばかりか，社会的価値追求と本来の目的を喪失しているという指摘が提起されると，これを踏まえ，政府は2013年に「二次社会的企業育成基本計画」を稼動させた。それを基に開催された「第二次5か年計画」（2013～2017）は，社会的企業の持続可能性の強化とオーダーメイド型の支援制度の確立，社会的企業の役割の拡大と成果の普及，民間企業と地域社会の協力関係の強化を主な政策の方向とし，2017年までに社会的企業数を3千社に拡大する計画である。第一に，社会的企業の持続可能性の強化のための取り組みとしては，社会的企業の市場開拓の支援と財政的支援・投資の拡大，社会的企業からなる公共調達の拡大，そして，助成金制度の改善が含まれ，第二に，オーダーメイド

型支援制度の確立に向けては，コンサルティング・サービスの拡大と効率化，支援組織の能力およびインフラの強化，社会的企業家に対する研修の拡大，そして，継続的サービスの提供のための具体的な取り組みが展開される。第三に，社会的企業の役割の拡大と成果の普及を実現するため，社会的企業の役割の拡大と責任の強化，成功モデルの普及，そして，社会的企業に関する合意の普及に関する取り組みが行われ，第四に，民間企業と地域社会の協力関係強化のためには，社会的企業に対する民間企業の支援強化，民間部門の人的資源と社会的企業とのつながりの強化，社会的企業と関連地域社会や産業との交流推進の促進と関連し，「一企業一社会的企業キャンペイン」，および引退した専門職や壮年期の失業者の社会的企業への就業促進などを含む活動が試みられる[5]。

5-2. 単一政策目標による法律構成から包括的な法律枠組みへ

　韓国の制度化を通じた社会的企業の育成は，社会的企業の生み出す価値を雇用の提供という狭義に限定し，実際に，韓国の社会的企業は，市場競争に弱い低付加価値の単純労働型が多く，短期的，低賃金の雇用水準に留まっており，それが社会的企業の提供する商品とサービスの質の低下に影響を及ぼすと指摘される。その背景としては，韓国の社会的企業が，経済危機以降の経済回復と新たな雇用の創出を模索する過程から出発し，立法過程において，脆弱階層の雇用の拡大といった短期間における政策的実績を優先にし，更に，その主導権が雇用労働部に握られた結果，雇用創出といった点に自然に政策目標の中心が偏ったことが挙げられる。

　その問題意識から出発し，近日，韓国政府は，社会的経済基本法の制定に取り掛かっている。社会的経済基本法は雇用労働部総括下の社会的企業を中心に，社会的経済の主体として活動する共同組合（企画財政部），自活企業（保険福祉部），マウル企業（行政自治部），農業村共同体会社（農林畜産食品部）を企画財政部の総括部署に統一し，政策総括・調整を行い，社会的企業振興院を中央自活センターとともに社会的経済院として改造する包括的な仕組みを取る。それにより，各部署が競争的に社会的経済組織を養成することによる，政府の財政無駄と事業成果の不備，政府からの行政力の無駄と民間からの過度な重複的行

政の規制の発生，社会的経済組織間の不必要な競争による収益性の悪化，部署間の重複投資による支援事業の不実化，社会的経済主体間のシナジー創出の障害などの問題を解決する。究極的には，政府の支援・伝達体系の効率化による政策効果を向上し，「ともに豊かな社会」への社会価値の転換と国家福祉財政負担の少ない福祉体系の構築することが狙われる[6]。

　一方，社会的経済基本法に関しては，それが，社会的経済の主体の必要によって要求され，準備された法律ではないため，市民自立性の侵害と市民社会主導性の弱化という否定的結果を招く恐れがあることから，賛否の議論が残っており，当法律の制定においては，社会的経済活動主体の能動性と主導性が問題の核心になると思われる。

5-3. 政府依存型から民間参与型の財務構造への移転

　韓国の社会的企業は，事業性が弱く，政府の助成金への依存度が高い上，財政状況の劣悪で，小規模にとどまっていることが特徴としてみられる。韓国国会立法調査処（2014）によると，2011年度の韓国社会的企業の売上はGDPの0.04％，全体雇用の0.1％を占め，2006年度，英国の社会的企業の売上がGDPの2％，全体雇用の5％（OECD平均4.4％）を占めることに比べ，とても小さいことが分かる。2008年度の社会的企業の事業費構成は，総事業費の中，政府支援金が27.8％と高く，それに対し民間機関による事業費は全体の2.2％に過ぎない（雇用労働部, 2010）。特に，現支援体制下では，政府の一時的な人件費などの支援が終了した後の社会的企業の生存に対する支援策が不十分であり，一時的期間に自立することができない場合には，廃業の危機を余儀なくされる支援体制の構造的問題を抱える。国会立法調査処（2014）の調査からは，「人件費の支援が中断されると58.6％の企業が廃業したり，一般企業に転換し，人員削減をすることになり，19.5％は脆弱階層勤労者を一般勤労者に交替する」と説明する。しかし，納入者の負担で支えられる公共政策は，予算の不足，効率性の低迷などの限界に至る可能性が高く，また，社会的費用の増大と市場の失敗にも繋がる恐れがある。

　これらを問題意識に，韓国政府は社会的企業の育成支援の直接支援から間接

支援への転換を展開してきた。これまでの試みとしては，大きく政府による貸出と信用保証，ファンド型の支援，マイクロファイナンス，そして，社会的基金の設立が挙げられる。第一に，社会的企業に対する公的金融支援は，主に政府による貸出と信用保証によって実現される。その中，雇用労働部を中心に事業運営などに必要な資金を長期・低利に貸し付ける場合が最も多く，信用保証の場合には，社会的企業を対する特別保証が提供される。第二に，政府は，民間の社会的企業に対する投資を活性化させるため，2011年には，公共ファンド型の投資インフラ，社会的企業投資組合を結成し，政府（雇用労働部）は毎年25億ウォンのシードマネーを出資している。また，産業銀行は2011年に10億ウォン規模のソーシャルファンド（社会問題を改善するなど，公益的目的に基づく社会的ファンド）を打ち上げ，多様な金融手段の提供などを含め社会の脆弱階層を支援している。第三に，マイクロファイナンスが挙げられる。2009年には信用不良者（信用等級7級以下）の自活に必要な創業資金，運営資金などを無担保・無保証で支援する「微小金融中央財団」を設立し，低金利融資（マイクロクレジット）事業を展開している。また，営利マイクロクレジット機関として2002年に設立を果たした社会連帯銀行は，2011年から公的資金および民間資本の社会的投資を実現するSFPP（Social Finance Pilot Project）を展開し，社会的に影響力のある社会的企業を支援する。第四に，社会的基金に関して，韓国では，2012年7月にアジアでは最初の社会投資基金として「ソウル特別市社会投資基金」[7]が制定・実施され，2016年末までを基金存続期限とし，主に融資事業と投資事業を展開している。また，2013年には，低所得家庭の実質的な金融負担を軽減し，金融市場の安定と経済活性化を実現することを目的に，「国民幸福基金」を設立し，連帯債権の債務調整，高金利貸付の低金利への転換貸付および自活プログラムの提供など，債務者に対する少額貸付を展開している[8]。

　ただ，韓国の社会的企業に対する資本調達は，以前として政府主導下で展開され，財源も政府と企業に依存しており，市民による出資があまりみられないなどから，まだ初期段階であると言える。更に，韓国の社会的企業の多くが，組織形態と規模，財務条件の側面からは，資本市場を通じて資金調達が困難で

あることが事実であり[9]、経済的目的と共に社会的目的を共に実現する社会的企業の特徴を顧慮した上で、既存資本市場のメカニズムとは異なる財源調達方法を確保することが求められる[10]。

5-4. 中央集中から地域分散・地域協力の強化

　最近は、多くの地方自治体は地方経済の回復を目指し、社会的企業の育成に積極的に取り組むことを示している。社会的企業の全国的な拡散と地域別の発展の格差を縮める上で貢献している。その中でも、ソウル市は、社会的経済のGDP比2％、雇用全体比8％を占めることを目標に、2010年にソウル型の社会的企業を育成することを示し、2011年にソウル市長に就任した朴元淳によって、社会的企業に対する支援政策が見直され、そのパラダイム・シフトに関する構想が打ち上げられた。ソウル型社会的企業とは、社会的サービスを提供する目的をもって活動する企業（団体）のうち、雇用労働部の認証条件は満たしていないが、社会的目的の実現と収益性の創出の成長可能性が高い、潜在的社会的企業と定義することが出来る。ソウル型社会的企業は、既存の社会的企業を通す脆弱階層の職業層創出が、政府主導に短期的かつ量的成長を重視する上で推進されてきたことから、市民主導の社会革新体の育成と社会的経済の活性化（長期性、雇用・収益は結果物）を目指し、そのための支援戦略として、既存の限時的人件費の支援（画一性、衡平性原則）から、事業費・市場造成・社会投資基金に及ぶ体系的な中間支援システムを構築と成長段階に合わせた総合支援体制の構築を強調する。2012年には、社会的企業、協同組合、マウル企業などを包括的に育成支援する「社会的経済課」を新設し、社会的企業に対する公共調達の実績は2012年の440億ウォンから翌年には622億ウォンにまで増加した[11]。

6. 社会的企業の育成のための戦略と課題

　韓国の社会的企業の育成のための経験に基づくと、社会的企業の育成においては、以下の考察すべき課題を提示することができる。

第一に，包括的かつ柔軟な法制度を具備する必要がある。日本型の「社会的事業法人」の構想もあるものの，関連法制度の整備では韓国が先行している。韓国の経験から，社会的企業の発展のための法制度の整備の上では，①社会的企業の範囲および対象について，それを制限するより包括的な枠組みを取り既存の第三セクターとも組み合い，多様な社会的価値を効率的に達成すること，②社会的企業の育成を国政課題の戦略的管理として捉え，関連政策を管轄する単一総括部署を最高政策機構の直属常説委員会として設置し，社会経済体制の構築という政策目標を効果的に達成すること，③政策差別化を通じ，企業の生涯周期によって求められる多様な経営支援，販路支援，金融支援することが重要であろう。

第二に，公共市場の拡大と社会的金融システムを構築する必要である。政府主導の制度化による促進が齎す肯定的な効果が明らかであるが，過当な政府規制による社会的企業の展開は，市場経済の機能を歪曲し，社会的企業の多角化を妨げる可能性がある。社会的企業の発展早期における政府による直接的な支援方法は，長期的には，民間主導の間接的な調達方式に展開させ，経営の成果に対し，多様な資金供給主体による評価を受け，また，需要と供給市場を共に拡大していく必要がある。そのためには資本市場を活用した資本調達体制の構築が重要であり，その実現を向け，①社会的企業育成を社会経済関連政策と連携し，位置づけること，②社会的価値を生み出す社会的企業，もしくはプロジェクトを支援する社会的金融（Social Finance）体制の普及に励むと同時に，社会的企業に対する民間の投資を活性化するため，関連したガイドラインと適当な税制恵沢を整備すること，③共済制度の導入し，社会的企業の自立のための金融環境の成熟していない段階において，社会的企業の危機管理能力と財政の独立性，および健全性を高めることができよう。

第三には，中間支援体制の開発と育成が挙げられる。社会的企業の生態系の健康な成長のためには，社会的企業を支援する多様な組織の発展が必要であるが，中間支援期間は市場—政府および自治体—社会的企業間の実質的なコーディネーターとして，社会的企業の生産と販売活動に必要な資金，人材，営業などの経営資源を社会的企業を結びつけ，特に社会的企業の関連インフラの脆

弱な発展初期において，その設立と運営，活性化を支援する上で重要な役割を果たす。中間支援組織自らは，革新と人材支援に対する投資を通じ，自らの支援能力を強め，地方資源を引き入れるなど，政府支援予算以外の収益源の多様化することが重要であり，政府は，市域政府と協力し，関連政策を制定するなど，地域における中間支援組織の事業の規模化，専門化，システム化，シナジー化，ネットワーク化を促進することが出来よう。

(1)「社会的企業育成法施行令」では，脆弱階層を，世帯月平均所得が全国世帯平均所得の100分の60以下の者などと定めている。
(2)「社会的企業育成法」第2条では，連携企業を，特定の社会的企業に対し財政支援，経営諮問など多様な支援を行う企業であって，その社会的企業と人的，物的，法的に独立している者と定義する。
(3) 2012年に雇用労働部が発表した資料によると，社会的企業関連予算1,760億ウォンの中，人件費が占める割合や70％と最も高い。
(4) 詳しくは韓国社会的企業振興院ホームページ，http://www.socialenterprise.or.kr を参照。
(5) 独立行政法人労働政策研究・研修機構ホームページから接続，
http://www.jil.go.jp/foreign/jihou/2013_2/korea_01.html。
(6) 2013年12月に与党（セヌリ党）の最高委員会で社会的経済特別委員会の構成が議決され，2014年4月には社会的経済基本法案が発議された。当法案に関しては，ソウル社会的経済ポータルから接続，「『社会的経済基本法』制定のための公聴会」(2014)（http://sehub.net/archives/6410）を参考することができる。
(7) 詳しくは財団法人韓国社会投資ホームページ http://www.social-investment.kr およびソウル特別市ホームページから接続，「ソウル特別市社会投資基金の設置および運用に関する条例」(http://opengov.seoul.go.kr/sanction/1389686) を参照。
(8) 詳しくは国民幸福基金ホームページ http://www.happyfund.or.kr と韓国資産管理工事ホームページ http://www.kamco.or.kr から参照。
(9) 고용노동부 (2013) によると，韓国の社会的企業の中，53％は株式と社債の発行を通じた資金調達が法的に禁じられている企業形態（民法上の法人，非営利民間団体）であり，また，社会的企業の平均資産規模の5.20億ウォンの中，平均負債は2.91億ウォン，平均資本は2.30億ウォンと，設立当時から資本規模が零細し，財政構造が脆弱な上，収入構造も劣悪なため，有価証券市場およびコスダック市場の上場要件を満たす社会的企業はほとんど存在しない。
(10) 各支援策の具体的な内容と運用，これまでの実績などに関しては，韓国雇用労働部による「成果連携型社会的企業資本市場構築法案研究」(2013) を参考することができる。

(11) ソウル型社会的企業に関するより詳しい情報は，ソウル特別市庁ホームページから接続，「ソウル型社会的企業」（http://se.seoul.go.kr/index.action）で参照することができる。

＜参考文献＞
柏井宏之（2013）「社会的企業の展開―日韓市民交流とその比較」，『大原社会問題研究所雑誌』，No. 662, pp. 28-47。
韓国国民幸福基金。Available at http://www.happyfund.or.kr, Accessed 2015 年 4 月 19 日。
韓国雇用労働部。Available at http://www.moel.go.kr, Accessed 2015 年 4 月 23 日。
韓国資産管理工事。Available at http://www.kamco.or.kr, Accessed 2015 年 4 月 19 日。
韓国社会的企業振興院。Available at http://www.socialenterprise.or.kr, Accessed 2015 年 4 月 20 日。
桔川純子（2010）「韓国市民運動の新しい展開―「社会的企業育成法」成立の背景」，『大阪経済法科大学アジア太平洋研究センター』，第 8 号，pp. 3-9。
金早雪（2011）「韓国の青年雇用ミスマッチへの 2 つの対応戦略―技術教育改革と社会貢献・起業支援」，『海外社会保障研究』，No. 176, pp. 53-65。
姜美羅，落合俊郎（2011）「韓国の社会的企業の現状と課題」，『広島大学大学院教育学研究科附属特別支援教育実践センター研究紀要』，第 9 号, pp. 39-50。
小関隆志（2015.03.14）「韓国における社会的企業育成とマイクロファイナンス」日本 NPO 学会第 17 回年次大会。Available at http://www.kisc.meiji.ac.jp/~koseki/results/JANPORA 17-panel.pdf, Accessed 2015 年 4 月 17 日。
駒崎ナエコ，小倉綾乃（2013）「韓国における社会的企業の展開―背景，事例，課題」，『損保ジャパン総研レポート』，第 63 号，pp. 25-41。
白井京（2008）「韓国における格差問題への対応―非正規職保護法と社会的企業育成法」，『国立国会図書館調査及び立法考察局』，外国の立法 No. 236, pp. 123-135。
ソウル市庁。Available at http://www.seoul.go.kr, Accessed 2015 年 4 月 25 日。
ソウル社会的経済ポータル。Available at http://sehub.net, Accessed 2015 年 4 月 26 日。
塚本一郎（2012）「労働統合と社会的企業」，全労済協会「生活保障研究会」，第 11 回研究。Available at http://zenrosaikyokai.or.jp/think_tank/action/report/lifegua/pdf/seikatsu_11_t.pdf, Accessed 2015 年 4 月 18 日。
独立行政法人労働政策研究・研修機構。Available at http://www.jil.go.jp, Accessed 2015 年 4 月 20 日。
友岡有希（2014）「韓国における低所得層政策の歴史的展開とその現状」，『東亜経済研究』，第 73 巻第 1 号，pp. 57-76。
内閣府（2011）『社会的企業についての法人制度及び支援の在り方に関する海外現地調査』，内閣府。Available at http://www5.cao.go.jp/npc/pdf/syakaiteki-kaigai.pdf, Accessed 2015 年 3 月 15 日。

橋本 理（2011）「「労働統合型社会的企業」論の展開―韓国の事例から」,『関西大学社会学部紀要』, 第42巻第3号, pp. 83-102。
고용노동부 (2010)「사회적기업 육성을 위한 자본시장 연구Ⅱ」, 고용노동부。
――― (2013)「성과연계형 사회적기업 자본시장 구축방안 연구」, 자본시장연구원。
곽선화 (2011)『사회적기업의 사회적, 경제적 성과분석』, 한국사회적기업진흥원。
곽관훈 (2012)「사회적기업 육성법제 및 기타 창업관련 법제 정비방안」,『증권법연구』, 1-27。
김상희 (2009)「사회적기업의 지속 가능한 발전 방향」, 국정감사 정책보고서(2)。
김성기 (2009)「사회적기업 특성에 관한 쟁점과 함의」,『사회복지정책』, 36(2), 139-166。
김인선, 노희진 (2013)「사회적기업의 육성방안」,『경영사학』, 28(2), 61-90。
김혜원 (2009)「한국 사회적 기업 정책의 형성과 전망」,『동향과 전망』, 75, 74-108。
노희진, 조영복, 안수현 (2012)『사회적기업 육성을 위한 자본시장 조성방안 연구』, 자본시장연구원。
문보경 (2014)「사회적경제 기본법 제정의 주요 이슈 점검」,『자활읽기』, (12), 12-20。
박태정 (2011)「사회적기업육성법의 주요 쟁점과 개선방향에 관한 연구」,『인문사회과학연구』12(2), 61-88。
박찬임, 김혜원, 이재원 (2009)「사회적기업 시장 확대를 위한 지방자치단체 공공조달 연구」, 고용노동부。
손을춘 (2014)「사회적기업 지원제도의 문제점과 개선방안」, 대한민국국회。
윤지연 (2014)「팽창하는 '사회적기업', 목적 상실 … 정리해고 위험도―정부 - 여야가 추진하는 사회적기업 확대, 면피용으로 전락하나」, 민중언론 참세상（http://www.newscham.net/news/view.php?board=news&nid=73245）―2015.4.20。
이대영, 이상희 (2009)「한국 사회적기업 연구 동향에 대한 분석과 고찰」,『한국비영리연구』, 8(2), 187-217。
이정봉 (2010)「사회적기업의노동조건현황과과제」,『노동사회연구』, 3/4 (151), 58-64。
이주호 (2013)「협동조합기본법 제정과 사회적기업 환경변화 분석」,『사회적기업과 정책연구』, 3(1), 67-96。
이희수 (2011)「사회적기업 관련법제에 관한 비교연구」,『사회적기업연구』, 4(2), 50-87。
장원봉 (2008)「한국 사회적 기업의 실태와 전망」,『동향과 전망』, 75, 47-73。
조영복, 곽선화 옮김 (2011)『영국의 사회적기업 육성 계획』, 시그마프레스。
――― (2011)「한국의 사회적기업 육성모델과 국제비교」,『사회적기업연구』, 4(1), 81-105。
한국사회적기업진흥원 (2014)「사회적기업 개요집」, 한국사회적기업진흥원。

4　消費財の情報特性がCSR活動に与える影響の分析

吉田賢一

早稲田大学大学院商学研究科博士後期課程

【要旨】
本研究では，CSR活動を企業のレピュテーションを向上させる投資と考え，その前提のもと，消費者が品質を認識しにくい財・サービスを供給する企業ほど，より積極的にCSR活動を行うと推測した。企業が供給する消費財の情報特性がCSR活動の実施にどのような影響を与えるか，CSR活動の代理変数に東洋経済新報社が提供する第6回CSR企業ランキングを用いて，その関係性を計量分析により明らかにした。分析の結果からは，企業と消費者との間の情報の非対称性が相対的に大きい「経験財」，特に「耐久性を持つタイプの経験財」を供給する企業がよりCSR活動に積極的であることが確認された。

キーワード：CSR活動，消費財の情報特性，決定要因分析

1. はじめに

これまでの日本企業を対象としたCSR研究は，CSR活動が企業パフォーマンスに与える影響の分析に関心が集中しており，両者は正の相関を持つことが確認されている[1]。他方で企業パフォーマンス以外を対象とするCSR活動の決定要因分析は進んでいない。この現状に対して首藤（2012）はCSR活動が企業パフォーマンスと結び付く際の経路，そして，企業がCSR活動を行うインセンティブ構造の分析を行う必要性を提起した。

本研究は，消費財の情報特性（企業と消費者の間に存在する非対称情報の程度）がCSR活動に与える影響に注目し，これを実証分析により明らかとする。具

体的には，Siegel and Vitaliano (2007) に従い，(1) 財・サービスを購入前に財の価値を評価可能な探索財（衣料・家具など），(2) 購入後に財の価値を評価できる経験財のうち，購入頻度が多い経験財［耐久性なし］（食品など），(3) 購入頻度が少ない経験財［耐久性あり］（自動車など），(4) サービスとして供給される経験財［サービス］（インターネットサービスなど），そして，(5) 購入後も財の価値を評価できない信頼財（投資助言など）の5タイプに財・サービスを分類してCSR活動への寄与度の違いを分析する。本稿の目的は，この検証を通じて日本企業がCSR活動を行うインセンティブ構造を解明することである。

以降の本稿の構成は次の通りである。第2節では先行研究のレビューとして戦略的CSR理論の展開と日本企業を対象とした実証研究について述べる。第3節では先行研究をもとに仮説の設定を行う。第4節では推計式（基本推計）を導出する。第5節ではCSRデータと財・サービスの分類の仕方について概要を述べる。第6節では基本推計による分析の結果を報告する。第7節ではサンプルセレクションバイアスに対処した推計と，別の説明変数を用いた推計を行い，頑健性を検証する。第8節では結論を述べる。

2. 先行研究

2-1. 戦略的CSR理論

戦略的CSR理論を提唱する先駆けとなった研究の1つにMcWilliams and Siegel (2001) がある[2]。McWilliams and Siegel (2001) はCSR活動を企業経営者の合理的な意思決定により，長期的な企業価値の最大化を達成することを目的に実施される投資と見なした。また，購入前に財の価値を評価可能な「探索財」を供給する企業よりも，購入後に財の価値を評価できる「経験財」を供給する企業の方がより積極的にCSR活動に取り組むとの仮説も提示した。

この戦略的CSR理論をベースとして，経営者がCSR活動へ投資を行う際のインセンティブメカニズムをAoki (2010) では以下のように説明した。

企業がCSR活動を実施した場合，CSR活動によるコストが価格へと移転さ

れることを厭わない，社会的な意識の高いステークホルダーの存在によって(1) 長期的な売上高や利潤への貢献，(2) 社会関係資本のダメージの回避，さらには，社会関係資本の蓄積への貢献が見込まれる。このとき当該企業が蓄積する社会関係資本は，社会的な意識の高いステークホルダーに正のシグナルの発信として認知されるため，企業はCSR活動へのコストを差し引いて余りある純長期利潤の獲得という展望を持つことが可能となる。このような長期的な利潤の獲得というインセンティブが，経営者による戦略的なCSR活動を推進していく際の動機づけになるとした。

Bénabou and Tirole (2010) は，CSR活動には (1) 長期的な企業価値に寄与する活動，(2) 社会貢献活動に関心を持つステークホルダーを代行する活動，そして，(3) 企業の内部者 (とりわけ経営者) の自己意識を満たすための活動，という3つの解釈可能な視点が存在することを指摘し，(1) の視点は戦略的CSR理論の文脈と一致する解釈であり，(3) の視点はFriedman (1970) に代表される，CSR活動を経営者によるエージェンシーコストと見なす見解と一致する解釈であると整理した。

Siegel and Vitaliano (2007) では，McWilliams and Siegel (2001) が提示した，CSR活動に関する経営判断はその企業が供給する財・サービスのタイプによって決まるという仮説を実証分析により検証した。分析からは (1) 経験財，信頼財を供給する企業はCSR活動に積極的に取り組む企業であり，(2) 探索財を供給する企業はCSR活動に消極的な企業であることが示された。この結果はMcWilliams and Siegel (2001) で示された仮説と整合的なものであった。

2-2. 日本企業を対象とした実証研究

日本での先駆的な研究にはTanimoto and Suzuki (2005) と首藤他 (2006) がある。Tanimoto and Suzuki (2005) は企業規模が大きく，環境に関連した事業を行っていて，外国人保有比率が高く，海外売上高が高いほどGRIガイドラインを採用する傾向があることを報告した。首藤他 (2006) では頑健性に欠けるとしながらも，(1) 明確なCSR方針を持つ企業は総花的なCSR活動を行っている企業よりも高いパフォーマンス傾向があること，(2) CSR活動へ

の取り組みは経営リスクを軽減する効果があること，(3) CSR 活動に積極的な企業は株式市場でポジティブに評価されることを明らかとした。

さらに首藤・竹原（2008a, 2008b）では，CSR 活動と強い関連を持つ，企業規模，産業特性，成熟度をコントロールしてもなお，CSR 活動に積極的な企業は成長性や市場評価の面でパフォーマンスの高い企業であることを報告した。また，非財務情報開示に積極的な企業や，消費者および地域とのコミュニケーションを重視している企業は，市場評価が安定する傾向を持つことも合わせて指摘した。

続いて行われた Suto and Takehara (2013) では，東洋経済新報社が保有する CSR データを用いて，日本企業を対象とした CSR 活動と企業パフォーマンスとの関係を分析した。同研究の主たる貢献は，従来の会計指標や市場評価指標（トービンの Q）に加え，ジェンセンの a などの株価リターンを基とした企業パフォーマンス指標を導入したことにある。CSR 指標と企業パフォーマンスの内生性を考慮した（二段階最小二乗法を用いた）推計を行っても，CSR 活動と企業の収益性との間には正の相関関係が存在すること，そして，CSR 活動が企業の経営リスクを引下げる効果を持つことを示した。

一方で，CSR 活動と企業パフォーマンスとの関係に負の相関関係を指摘する日本企業を対象とした実証研究も存在する[3]。広田（2012）は，「株主以外のステークホルダーの非金銭的な価値」と会計利益率，株価水準，株式投資収益率との間の相関関係について分析を行った。東洋経済新報社「第 5 回 CSR 企業ランキング」の雇用，環境，社会性といった CSR 項目を「株主以外のステークホルダーの非金銭的な価値」の代理指標とし，「株主以外のステークホルダーの非金銭的な価値」と会計利益率，株価水準，株式投資収益率との間には負の相関関係があることを報告した。

このような日本企業を対象とする CSR 研究が企業パフォーマンスとの関係の特定という関心に焦点が集中している現状に対して，首藤（2012）では欧米での CSR 研究と大きな隔たりがあることを指摘し，これまで関心が向けられてこなかった CSR 活動と企業パフォーマンスとの間にある「経路」，「メカニズム」，「インセンティブ」の考察が進められるべきであると問題提議がなされた。

3. 仮説

　McWilliams and Siegel（2001）を参考とし，消費財産業の情報特性が日本企業のCSR活動の実施に与える影響についての仮説を設定する[4]。社会的な活動への関心の高いステークホルダーの存在を所与の条件としたときに，当該企業にとっては，そのようなステークホルダーに訴求力のあるCSR活動に従事することがある種の製品差別化戦略となる。

　ここでは，そのようなステークホルダーを消費者と仮定する。経営者がCSR活動に投資決定を行う際には（1）自社の供給する財・サービスに関する情報の非対称性の程度，（2）供給主体の信頼性が消費者の財・サービスへの選好に影響を与える度合いが重要な判断要素となる。

　消費者は価値を十分に知ることが難しい（相対的に情報の非対称性が大きい）財を購入する際，供給企業の信頼性を判断基準とする。本研究ではCSR活動を企業のレピュテーションを向上させる投資と捉える。このとき，財・サービスについて情報の非対称性が大きな企業ほど，企業自身の信頼性を担保する手段として，より積極的にCSR活動を行うことが予想される。

　Siegel and Vitaliano（2007）では（1）探索財，（2）経験財［耐久性なし］，（3）経験財［耐久性あり］，（4）経験財［サービス］，（5）信頼財の5タイプに分類を行っている。本稿では，このうちで特に情報の非対称性に際立った違いを持つ，探索財と経験財，耐久性の低い経験財と耐久性の高い経験財に注目し，消費財産業の情報特性が日本企業のCSR活動の実施に与える影響について仮説を設定する。

　探索財は財の購入前であっても，その価値を容易に評価することができる財である。経験財は財を購入した後で財を使用（消費）しなければ，その価値を評価できない財である。したがって，探索財よりも経験財の方が相対的に情報の非対称性は大きくなり，供給主体の信頼性が買い手の財の選好に影響を与える度合いは高くなる。このような考察から以下の仮説を提示する。

仮説 1　探索財よりも経験財を供給する企業の方がより CSR 活動に積極的な企業である。

経験財はさらに，財の耐久性によって経験財［耐久性なし］と経験財［耐久性あり］に分けることができる。経験財［耐久性なし］は頻繁に購入することができ，購入の都度，財の価値判断が可能な経験財である。経験財［耐久性あり］は一定の期間継続して消費することが可能で，一般には頻繁に購入されることのない財であり，財の品質を十分に理解するのに長い期間が必要となる。したがって，経験財［耐久性なし］よりも経験財［耐久性あり］の方が，相対的に情報の非対称性は大きくなり，供給主体の信頼性が当該企業の供給する財の選好に影響を与える度合いも高くなる。このような考察から以下の仮説を提示する。

仮説 2　経験財のなかでも，耐久性の低い財よりも耐久性の高い財を供給する企業の方がより CSR 活動に積極的な企業である。

企業経営者による戦略的な意思決定のもとで，日本企業の CSR 活動が実施されているならば，供給する財・サービスの特性が CSR 活動の積極性に影響を与えるはずである。換言すれば，戦略的 CSR 理論のもとでは，探索財を供給する企業よりも，経験財を供給する企業の方がより積極的に CSR 活動に取り組み，経験財［耐久性なし］を供給する企業よりも，経験財［耐久性あり］を供給する企業の方がより積極的に CSR 活動に取り組んでいると推測される。

以下では，消費財産業の商品特性が日本企業の CSR 活動に与える影響について計量分析により明らかにしていく。

4. 推計式（基本推計）

Siegel and Vitaliano（2007）を参考にして，消費財産業の商品特性が日本企業の CSR 活動の実施に与える影響を分析するクロスセクション回帰分析を行う。

$$CSR_{it} = a_0 + a_1 GOODTYPE_{it-1} + a_2 ASS_{it-1} + a_3 RD_{it-1} + a_4 ADV_{it-1} + \varepsilon \quad (1)$$

添字 i は企業 i を，添字 t は時点 t を示す。被説明変数は CSR データであり，説明変数は 5 種類の財・サービスのタイプを表す代理変数である。コントロール変数は企業の規模，技術的競争力，製品差別化を用いた。

被説明変数の CSR データには東洋経済新報社の提供する（1）CSR スコア，CSR スコアの構成項目である（2）雇用スコア，（3）環境スコア，（4）企業統治＋社会性スコアを採用した。説明変数の GOODTYPE は Siegel and Vitaliano（2007）に従い，5 種類の財・サービスのタイプを用いた。5 種類の財・サービスとは（1）探索財，（2）経験財［耐久性なし］，（3）経験財［耐久性あり］，（4）経験財［サービス］，そして，（5）信頼財のことである。

なお，本研究では財・サービスのタイプを 2 つの異なる方法によって数値化した。このように異なる数値化の仕方を行った理由については後述する。この基本推計では，分類 I を用いて分析した。

ASS，RD，ADV はコントロール変数であり，ε は誤差項を表している。ASS は McWilliams and Siegel（2000, 2001），および，首藤・竹原（2008a, 2008b）等に従い，企業規模を導入し，その代理変数は一期前の総資産を自然対数に変換した値を用いた。RD は企業の技術的競争力を導入し，その代理変数は一期前の研究開発投資集約度を使用した。そして，ADV は製品差別化を導入し，その代理変数は一期前の売上高広告費比率を使用した[5]。

技術的競争力と製品差別化の導入は，McWilliams and Siegel（2000, 2001）による理論を根拠としている。技術的競争力について，McWilliams and

表1　代理変数一覧

変数		代理変数		出所
被説明変数 CSR		CSRスコア	300点満点	東洋経済新報社 第6回CSR企業ランキング：第7回CSR調査に基づき作成
		雇用スコア	100点満点	
		環境スコア	100点満点	
		企業統治・社会性スコア	100点満点	
説明変数 財・サービスタイプ		探索財	[分類Ⅰ] ・日本標準産業分類（12回改定，小分類）を左記の5タイプに分類。 ・分類可能なセグメントが，企業の売上高に占める割合を数値化。	[分類Ⅰ] 企業財務データバンク，日経テレコン21
		経験財［耐久性なし］		
		経験財［耐久性あり］		
		経験財［サービス］	[分類Ⅱ] ・日経NEEDS業種分類（中分類67種）を左記の5タイプに分類。 ・該当する業種の場合は1となるダミー変数。	[分類Ⅱ] 日経ValueSearch
		信頼財		
規模	総資産	総資産対数値		NEEDS-Cges
技術的競争力	R&D集約度	R&D投資額／売上高		NEEDS-Cges，企業財務データバンク
製品差別化	売上高広告費比率	広告宣伝費／売上高		NEEDS-Cges，企業財務データバンク
収益性	ROA	業種等調整ROA		NEEDS-Cges
成熟度	創業年数	2010 − 設立年度		QUICK-Astra Manager
財務安全性	負債比率	（負債合計／総資産）×100		NEEDS-Cges
外部圧力	外国人持株比率	外国人保有比率		NEEDS-Cges

Siegel（2000, 2001）ではCSR活動の一部と見なすことが出来る，社会性を生み出す研究開発が存在することが指摘されており，この両者は正の相関関係を持つと予想されている。また，製品差別化について，McWilliams and Siegel（2000, 2001）では広告宣伝費を社会性の高い財・サービスの購入に関心の高い消費者に自社の取り組むCSR活動を認知させ，製品差別化をする手段として捉えており，この両者は正の相関関係を持つと予想されている。

以上に述べた代理変数の一覧は表1にまとめている。なお，本稿での仮説から予想される財・サービスの係数の符号は探索財がマイナス，経験財（経験財［耐久性なし］と経験財［耐久性あり］）がプラスである。そして，購入頻度が多い経験財［耐久性なし］と購入頻度が少ない経験財［耐久性あり］では，購入頻度が少ない経験財［耐久性あり］の方がより，係数がプラス方向に大きくとなることが予想される。

5. データ

5-1. CSR データ

CSR データは東洋経済新報社「第6回 CSR 企業ランキング」を用いた。「第6回 CSR 企業ランキング」は，「第7回 CSR 調査」（2011年6月実施）[6]と東洋経済新報社が保有する上場企業財務データに基づいて作成されている。このランキングの対象からは，銀行，証券，保険，その他金融が除かれている。

「第6回 CSR 企業ランキング」は CSR 評価（300点満点）と財務評価（300点満点）の2項目に分類され，合計600点満点で評価される。CSR 評価の項目は（1）雇用（100点満点），（2）環境（100点満点），（3）企業統治＋社会性（合計で100点満点）の合計300点満点で評価される。一方，財務評価の項目は（1）収益性（100点満点），（2）安全性（100点満点），（3）規模（100点満点）の計300点満点で評価される。

本研究では以下の方法でサンプルを抽出した。まず，「第6回 CSR 企業ランキング」で別掲されている金融機関30社を700社に加えて，計730社のランキングを作成し，このランキング（730社）から財務評価の項目を除き，CSR 項目のみによる数値評価を算出した上で，再び順位づけをする操作を行った。この一連の操作によって，銀行，証券，保険，その他金融などの金融機関を含む，730社の CSR スコア，雇用スコア，環境スコア，企業統治＋社会性スコアを得た。

5-2. 財・サービスの分類

本研究では，表2に示したSiegel and Vitaliano (2007) での財・サービスの分類に倣い，独自に分類表を作成し，その基準に則って財・サービスの分類作業を行った。本研究では，以下の2つの異なった方法（分類Iと分類II）を用いて分類した。なお，ここでの分類作業では，Siegel and Vitaliano (2007) に従い，消費財産業だけを分類の対象とした[7]。

5-2-1. 分類I

日本政策投資銀行の企業財務データバンクと，日経テレコン21（2013年5月15日時点）からセグメント情報を入手し，日本標準産業分類（12回改定）の小分類基準に則って，5つの財・サービスのタイプへと分類した[8]。ここでの分類の基準は表3にまとめた通りである[9]。

分類Iの財・サービスの分類作業の具体例をA社，B社，C社という架空の

表2 Siegel and Vitaliano (2007) での財・サービスのタイプ分類

探索財	経験財[耐久性なし]	経験財[耐久性あり]	経験財[サービス]	信頼財
衣類	健康・美容	住宅	広告	資産運用
家具	煙草	自動車	運送	信託
はきもの類	食品	電気器具	リゾート	資産運用管理
カーペット	クリーニング	ハードウェア	教育	ミューチュアルファンド
マットレス	新聞	薬品	トレーニング	保険
	事務用消耗品	眼鏡	旅行	ヘルスケア
		ソフトウェア	銀行	ウェイトコントロール
		看板	レンタカー	自動車修理
		書籍	エンターテイメント	
		スポーツ用品	ダイレクトメール	
		玩具	不動産	
		公益事業	貨物運搬	
			就職斡旋	
			情報（報道）	
			高齢者福祉施設	
			スポーツクラブ	
			宿泊施設	
			ゴミ収集	
			造園	

出所：Siegel and Vitaliano (2007), p. 780.
原資料：Nelson (1974), Liebermann and Flint-Goor (1996).

表3　本稿での財・サービスのタイプ分類 [分類 I]

探索財		経験財 [耐久性なし]		経験財 [耐久性あり]		経験財 [サービス]		信頼財	
外衣・シャツ製造業	116	畜産食料品製造業	091	建築工事業（住宅）	064	移動電気通信業	372	銀行	622
和装製品・その他の衣服・繊維製身の回り品製造業	118	水産食料品製造業	092	木造建築工事業（住宅）	065	有線放送業	383	クレジットカード業，割賦金融業	643
		野菜缶詰・果実缶詰・農産保存食料品製造業	093	医薬品製造業	165	インターネット附随サービス業	401	金融商品取引業	651
				民生用電気機械器具製造業	293	映像情報制作・配給業	411	商品先物取引業・商品投資業	652
家具製造業	131								
革製履物製造業	204	調味料製造業	094	電球・電気照明器具製造業	294	一般乗用旅客自動車運送業	432	生命保険業	671
		精穀・製粉業	096					損害保険業	672
洋食器・刃物・手道具・金物類製造業	242	パン・菓子製造業	097	通信機械器具・同関連機械器具製造業	301	集配利用運送業	444	結婚相談業（他に分類されない生活関連サービス業）	799
		動植物油脂製造業	098			航空運送業	461		
貴金属・宝石製品製造業	321	その他の食料品製造業	099	映像・音響機械器具製造業	302	百貨店，総合スーパー	561	職業紹介業	911
		清涼飲料製造業	101			その他の各種商品小売業	569		
		酒類製造業	102	電子計算機・同附属装置製造業	303	自動車小売業	591		
		茶・コーヒー製造業	103			通信販売・訪問販売小売業	611		
		たばこ製造業	105	自動車・同附属品製造業	311				
		下着類製造業	117	時計・同部分品製造業	323	不動産賃貸業	691		
		紙製品製造業	144			自動車賃貸業	704		
		化粧品・歯磨・その他の化粧用調整品製造業	166	楽器製造業	324	旅館・ホテル	751		
				がん具・運動用具製造業	325	食堂・レストラン	761		
		ペン・鉛筆・絵画用品・その他の事務用品製造業	326	他に分類されない製造業	329	旅行業	791		
						スポーツ施設提供業	804		
				出版業	414	公園・遊園地	805		
		新聞業	413			老人福祉・介護事業	854		

注：Siegel and Vitaliano (2007), p. 780を参考にして作成. 3桁の数字は日本標準産業分類（12回改定）の小分類を表す.

4 消費財の情報特性が CSR 活動に与える影響の分析　　305

表4　本稿での財・サービスのタイプ分類［分類Ⅱ］

探索財	経験財［耐久性なし］	経験財［耐久性あり］	経験財［サービス］	信頼財
衣料品・服装品	食品製造 飲料・たばこ・嗜好品 日用品・生活用品 弁当・デリバリー	家庭用電気機器 自動車 趣味・娯楽用品 バイオ・医薬品関連 医療・ヘルスケア・介護	不動産 各種商品小売 通信販売 飲食店 マスメディア コンテンツ制作・配信 レジャー・レジャー施設 生活関連サービス 教育	銀行 証券 保険 消費者・事業者金融

注：Siegel and Vitaliano (2007), p. 780を参考にして作成。日経 NEEDS 業種分類（中分類67種）のうちで分類可能であった業種を記載している。

企業を用いて以下で説明する。

A社は食品事業（40億円），医薬品事業（30億円），レストラン事業（30億円）の3つの事業セグメントを持つ売上高100億円の企業とする。A社の財・サービスの分類は，食品事業が経験財［耐久性なし］= 0.4，医薬品事業が経験財［耐久性あり］= 0.3，レストラン事業が経験財［サービス］= 0.3 と計算する。B社は衣服事業（50億円），化粧品事業（30億円），広告部門（20億円）の3つの事業セグメントを持つ売上高100億円の企業とする。B社の財・サービスの分類は，衣服事業が探索財 = 0.5，化粧品事業が経験財［耐久性あり］= 0.3 と計算する。広告事業は，ここでの財・サービスの分類の対象とはならずゼロの値とする。C社は化学繊維事業（50億円），医薬品原料事業（50億円）の2つの事業セグメントを持つ，売上高100億円の企業とする。C社の財・サービスの分類は，すべての事業セグメントが財・サービスの分類の対象とはならずゼロの値とする。

5-2-2. 分類Ⅱ

以上で述べた分類Ⅰは，企業が開示するセグメント情報をもとに日本標準産業分類の小分類基準に則って，分類作業を行っている。この作業プロセスには作業者による裁量の余地が大きく，恣意性を十分に排除することが難しい。このような問題を軽減させるため，本稿では，分類Ⅰとは異なった分類方法によ

る財・サービスのタイプ分けも行った。以下で述べる分類の仕方を分類IIとする。

分類IIでは日経 Value Search（2014年8月25日時点）からセグメント情報を入手し，日経 NEEDS 業種分類（中分類67種）に則って，5つの財・サービスのタイプへと分類した。分類IIの分類基準は表4にまとめた通りである。

分類IIは該当する業種の場合は1の値をとるダミー変数である。分類IIの財・サービスの分類作業の具体例を A 社，B 社という架空の企業を用いて以下で説明する。

A 社は「自動車」,「リース・レンタル」の2つの事業セグメントを持つ企業とする。「自動車」は経験財［耐久性あり］に該当するため1の値とし，「リース・レンタル」はここでの財・サービスの分類の対象とはならずゼロの値とする。B 社は「総合小売・食料品小売」,「飲食店」,「銀行」の3つの事業セグメントを持つ企業とする。「総合小売・食料品小売」,「飲食店」は経験財［サービス］に該当するため1の値とし，「銀行」は信頼財に該当するため1の値とする。

5-3. 基本統計量と相関係数行列

以上で説明した各変数の基本統計量は表5に記述した。表5のパネル B から E では，分類Iと分類IIの基準のもとで財のタイプ分けを行った際の該当企業の数と，該当企業の各 CSR スコアの傾向をまとめている。分類Iを基準とした分類での CSR スコア（表5のパネル B 上段）に注目すると，探索財，経験財［耐久性なし］，経験財［耐久性あり］に該当する企業は，それぞれ160.7点，189.2点，215.1点というスコアの平均値を示している。また，経験財［サービス］は169.5点，信頼財は230.7点であった。

分類IIを基準とした分類での CSR スコア（表5のパネル B 下段）に注目すると，探索財，経験財［耐久性なし］，経験財［耐久性あり］に該当する企業が CSR ランキングに入る確率は，それぞれ15.79％，22.31％，37.93％となっている。また，経験財［サービス］は12.65％，信頼財は44.44％であった。

表5のパネル F では，全サンプル企業と CSR ランキングに含まれた企業，

4 消費財の情報特性が CSR 活動に与える影響の分析　　307

それぞれについて，その他の変数の基本統計量をまとめた。この表からは，CSR ランキングに含まれた企業は，企業規模が大きく，R&D 集約度，収益性，安定性，外国人持株比率が高く，成熟期にある（創業年が古い）という特徴を持つことが読み取れる。

表5　基本統計量

パネル A　各種 CSR スコア　　(N=621)

	平均値	標準偏差	最小値	25%タイル値	中央値	75%タイル値	最大値
CSR スコア	177.2	50.8	88.9	135.3	168.9	216.9	296.3
雇用スコア	52.5	19.2	0.0	39.4	51.5	66.7	100.0
環境スコア	61.0	22.0	0.0	45.6	61.8	79.4	100.0
企業統治・社会性スコア	62.2	18.7	0.0	48.4	61.1	77.0	100.0

パネル B　財タイプ別　CSR スコア（上段：分類 I，下段：分類 II）

分類 I （N=621）

	該当社数	平均値	標準偏差	最小値	25%タイル値	中央値	75%タイル値	最大値
探索財	15	160.7	39.3	96.3	127.9	160.7	166.8	225.6
経験財[耐久性なし]	57	189.2	48.8	91.7	148.2	191.0	225.6	285.5
経験財[耐久性あり]	95	215.1	51.1	91.7	180.4	222.2	257.7	296.3
経験財[サービス]	91	169.5	46.8	96.6	136.6	157.8	197.8	296.3
信頼財	9	230.7	59.8	122.5	222.2	239.0	271.1	296.3

分類 II （N=3013）

	該当社数	rank in 率	平均値	標準偏差	最小値	25%タイル値	中央値	75%タイル値	最大値
探索財	57	15.79%	144.6	39.8	93.4	116.6	127.9	160.5	212.2
経験財[耐久性なし]	242	22.31%	192.5	46.5	113.7	151.6	189.5	225.6	285.5
経験財[耐久性あり]	261	37.93%	205.0	49.9	91.7	166.8	207.7	243.2	296.3
経験財[サービス]	830	12.65%	168.2	43.8	96.3	136.4	160.2	197.5	269.0
信頼財	9	44.44%	220.2	41.9	175.4	186.3	218.1	254.0	269.0

パネルC　財タイプ別　雇用スコア（上段：分類Ⅰ，下段：分類Ⅱ）

分類Ⅰ（N=621）

	該当社数	平均値	標準偏差	最小値	25%タイル値	中央値	75%タイル値	最大値
探索財	15	51.5	11.7	30.3	45.5	50.0	56.1	71.2
経験財[耐久性なし]	57	57.8	18.8	20.0	39.4	59.1	71.2	93.9
経験財[耐久性あり]	95	64.9	19.4	20.0	51.5	65.2	78.8	100.0
経験財[サービス]	91	53.5	19.7	0.0	39.4	54.5	68.2	100.0
信頼財	9	77.3	16.7	48.5	65.2	81.8	89.4	100.0

分類Ⅱ（N=3013）

	該当社数	rank in 率	平均値	標準偏差	最小値	25%タイル値	中央値	75%タイル値	最大値
探索財	57	15.79%	48.8	13.2	25.8	45.5	53.0	54.5	69.7
経験財[耐久性なし]	242	22.31%	59.3	17.6	28.8	43.9	59.1	71.2	93.9
経験財[耐久性あり]	261	37.93%	61.9	18.8	20.0	48.5	62.1	77.3	100.0
経験財[サービス]	830	12.65%	52.5	17.6	0.0	40.9	53.0	66.7	90.9
信頼財	9	44.44%	70.5	11.5	59.1	63.7	68.2	77.3	86.4

パネルD　財タイプ別　環境スコア（上段：分類Ⅰ，下段：分類Ⅱ）

分類Ⅰ（N=621）

	該当社数	平均値	標準偏差	最小値	25%タイル値	中央値	75%タイル値	最大値
探索財	15	50.5	23.4	20.0	22.1	52.9	69.1	85.3
経験財[耐久性なし]	57	63.7	18.5	20.0	51.5	64.7	80.9	95.6
経験財[耐久性あり]	95	74.5	19.8	20.0	63.2	80.9	88.2	98.5
経験財[サービス]	91	51.1	22.8	0.0	33.8	50.0	70.6	97.1
信頼財	9	74.0	28.1	20.0	66.2	82.4	94.1	98.5

分類Ⅱ (N=3013)

	該当社数	rank in 率	平均値	標準偏差	最小値	25%タイル値	中央値	75%タイル値	最大値
探索財	57	15.79%	38.9	19.7	20.0	20.0	38.2	50.0	69.1
経験財[耐久性なし]	242	22.31%	64.7	18.7	20.0	52.9	64.7	79.4	95.6
経験財[耐久性あり]	261	37.93%	71.8	19.2	20.0	58.8	76.5	86.8	98.5
経験財[サービス]	830	12.65%	51.9	23.1	0.0	33.8	52.9	72.1	100.0
信頼財	9	44.44%	66.2	21.9	38.2	49.3	69.1	83.1	88.2

パネルE 財タイプ別 企業統治・社会性スコア (上段:分類Ⅰ, 下段:分類Ⅱ)

分類Ⅰ (N=621)

	該当社数	平均値	標準偏差	最小値	25%タイル値	中央値	75%タイル値	最大値
探索財	15	58.6	11.6	40.5	48.4	57.9	70.6	77.0
経験財[耐久性なし]	57	67.7	15.7	34.9	58.7	66.7	80.2	96.0
経験財[耐久性あり]	95	75.7	17.5	34.1	62.7	81.0	89.7	99.2
経験財[サービス]	91	61.6	18.0	0.0	50.0	61.1	71.4	99.2
信頼財	9	79.5	19.0	45.2	74.6	80.2	94.4	99.2

分類Ⅱ (N=3013)

	該当社数	rank in 率	平均値	標準偏差	最小値	25%タイル値	中央値	75%タイル値	最大値
探索財	57	15.79%	57.0	13.2	40.5	47.6	57.9	61.9	77.8
経験財[耐久性なし]	242	22.31%	68.5	14.8	38.1	58.7	66.3	80.2	98.4
経験財[耐久性あり]	261	37.93%	71.3	17.5	34.9	57.9	73.0	84.9	100.0
経験財[サービス]	830	12.65%	61.0	17.1	0.0	50.8	61.1	70.6	94.4
信頼財	9	44.44%	83.5	12.2	69.0	73.4	85.4	93.7	94.4

パネルF　各種変数データ

全サンプル：3013社，rank in 企業：621社

		平均値	標準偏差	最小値	25%タイル値	中央値	75%タイル値	最大値
総資産	全サンプル	204.024	999.348	0.114	9.201	27.125	82.095	30349.290
(10億円)	rank in 企業	709.719	1996.679	2.453	44.203	136.991	468.178	30349.290
ln 総資産	全サンプル	10.320	1.741	4.736	9.127	10.208	11.316	17.228
	rank in 企業	11.935	1.756	7.805	10.697	11.828	13.057	17.228
R&D集約度	全サンプル	0.018	0.069	0.000	0.000	0.003	0.018	2.659
	rank in 企業	0.022	0.033	0.000	0.000	0.009	0.031	0.317
売上高広告費比率	全サンプル	0.009	0.027	0.000	0.000	0.000	0.002	0.426
	rank in 企業	0.008	0.021	0.000	0.000	0.000	0.004	0.165
ROA	全サンプル	−1.334	8.449	−88.447	−4.225	−1.458	1.592	116.938
	rank in 企業	0.147	5.728	−11.185	−2.486	−0.452	1.837	93.550
創業年数	全サンプル	51.472	24.824	1	33	58	66	129
	rank in 企業	62.781	22.714	3	52	62	75	129
負債比率	全サンプル	50.325	21.761	0.000	33.950	50.540	66.720	207.940
	rank in 企業	49.907	19.129	6.030	35.320	50.280	65.650	97.490
外国人持株比率	全サンプル	7.955	11.143	0.000	0.300	2.970	11.830	84.080
	rank in 企業	15.493	13.169	0.000	4.400	13.240	24.390	75.270

表6　相関係数行列

(N = 621)

	CSR	雇用	環境	企業統治・社会性	探索財	経験財[耐久性なし]	経験財[耐久性あり]	経験財[サービス]	信頼財	ln 総資産	R&D集約度	売上高広告費比率	ROA	創業年数	負債比率
雇用	0.848***														
環境	0.861***	0.605***													
企業統治・社会性	0.900***	0.791***	0.770***												
探索財	−0.053	−0.009	−0.082**	−0.027											
経験財[耐久性なし]	0.078*	0.091**	0.037	0.100**	−0.028										
経験財[耐久性あり]	0.288***	0.242***	0.250***	0.271***	−0.046	−0.090**									
経験財[サービス]	−0.069*	0.025	−0.202***	−0.021	−0.048	−0.086**	−0.101**								
信頼財	−0.006	0.028	−0.047	0.013	−0.008	−0.017	0.031	−0.008							
ln 総資産	0.781***	0.654***	0.681***	0.714***	−0.055	0.013	0.288***	0.039	−0.041***						
R&D集約度	0.191***	0.156***	0.195*	0.169***	−0.078	−0.065	0.439***	−0.204***	−0.018	0.135***					
売上高広告費比率	0.087***	0.146	−0.024***	0.122***	0.181	0.266***	0.254***	0.244***	0.007	0.059	0.054				
ROA	0.031	0.092	−0.033	0.066	0.023	0.036	0.046	0.096***	−0.058	0.033	0.048	0.260***			
創業年数	0.305***	0.222***	0.374***	0.280***	0.019	0.052	0.095***	−0.261***	−0.044	0.303***	0.100***	−0.176***	−0.136***		
負債比率	0.223***	0.143***	0.221***	0.194***	−0.064**	−0.123***	−0.017	0.071*	−0.062	0.336***	−0.256***	−0.114***	−0.238***	0.178***	
外国人持株比率	0.503***	0.479***	0.403***	0.455***	−0.028	0.002	0.310***	−0.013	−0.003*	0.614***	0.269***	0.160***	0.300***	0.092**	−0.086**

注：*** は1%，** は5%，* は10%水準で有意であることを示す。

表6は全変数間での相関係数の値を示している。環境スコアと探索財との間には負の相関が確認された一方で，各CSRスコアと経験財（経験財［耐久性なし］，経験財［耐久性あり］）との間にはおおむね正の相関がみられた。そして，CSRスコアと企業規模，R&D集約度，創業年数，外国人持株比率といった変数との間にも正の相関が確認された。

6. 推計結果

6-1. 基本推計
6-1-1. OLS 推定（N = 621）

分析の結果は表7のパネルAからDにまとめた通りである。パネルAはCSRスコアを，パネルBからDは雇用スコア，環境スコア，企業統治＋社会性スコアをそれぞれ被説明変数にして分析を行った結果である。各パネルの(1)から(5)式は，5つある財・サービスのタイプのうち，単独の財・サービスを説明変数とし，これにコントロール変数を加えてOLS推定を行った。そして，(6) 式は，すべての財・サービスのタイプを説明変数とし，コントロール変数とともにOLS推定を行った。

観察された事実は次の通りである。経験財［耐久性なし］の係数はおおむねプラスで有意であった。例えば，表7のパネルA，(6) 式では経験財［耐久性なし］の係数の値は11.876となっている。この係数の値は，企業が経験財［耐久性なし］に該当する財の供給をゼロから50%に増加させたとき，CSRスコアを5.9点上昇させるインパクトを持つことを意味する。

他方で，流通業や飲食チェーンを代表とする経験財［サービス］の係数はおおむねマイナスで有意であり，金融業を代表とする信頼財の係数は，被説明変数を環境スコアとした場合を除き，プラスで有意であった。また，企業規模，R&D集約度，売上高広告費比率においては，おおむねプラスで有意な係数の値がみられた。

経験財［耐久性なし］の係数がプラスで有意であったことは，仮説1と一部整合的である。だが，ここでの検証からは，経験財［耐久性なし］と経験財

表7 OLS推定 (N=621)

パネルA 被説明変数「CSRスコア」

説明変数	(1)	(2)	(3)	(4)	(5)	(6)
探索財	−4.607 (9.54)					−6.280 (9.75)
経験財 [耐久性なし]		14.889*** (4.94)				11.876** (5.21)
経験財 [耐久性あり]			5.686 (6.68)			5.015 (6.85)
経験財 [サービス]				−19.914*** (5.49)		−17.704*** (5.78)
信頼財					43.936*** (4.37)	41.792*** (5.57)
ln 総資産	22.199*** (0.66)	22.200*** (0.66)	22.035*** (0.70)	22.375*** (0.65)	22.248*** (0.66)	22.189*** (0.69)
R&D 集約度	129.268** (61.80)	139.174** (65.22)	113.476* (67.52)	95.822 (62.98)	131.181** (64.41)	90.030 (67.50)
売上高 広告費比率	97.189 (61.80)	46.072 (60.55)	76.636 (62.37)	155.637** (62.65)	91.318 (60.13)	104.812 (72.88)
_cons	−91.297*** (7.62)	−92.223*** (7.60)	−89.418*** (8.00)	−91.484*** (7.55)	−92.033*** (7.59)	−90.153*** (7.91)
観測数	621	621	621	621	621	621
adjusted R2	0.619	0.623	0.620	0.628	0.620	0.632

パネルB 被説明変数「雇用スコア」

説明変数	(1)	(2)	(3)	(4)	(5)	(6)
探索財	2.008 (3.47)					3.244 (3.63)
経験財 [耐久性なし]		5.192** (2.21)				5.660** (2.35)
経験財 [耐久性あり]			0.300 (2.62)			1.089 (2.70)
経験財 [サービス]				−1.036 (2.80)		0.284 (2.90)
信頼財					34.380*** (2.67)	35.046*** (3.15)
ln 総資産	6.983*** (0.33)	6.969*** (0.33)	6.966*** (0.34)	6.983*** (0.32)	6.999*** (0.32)	6.969*** (0.34)
R&D 集約度	37.687 (24.64)	40.071 (24.71)	36.187 (25.36)	35.282 (24.96)	37.522 (24.45)	38.948 (26.69)
売上高 広告費比率	94.789*** (24.21)	81.034*** (24.42)	96.239*** (24.62)	100.359*** (24.59)	96.515*** (23.24)	71.540** (29.85)
_cons	−32.474*** (3.90)	−32.577*** (3.90)	−32.238*** (4.08)	−32.347*** (3.89)	−32.708*** (3.85)	−32.746*** (4.05)
観測数	621	621	621	621	621	621
adjusted R2	0.444	0.447	0.443	0.444	0.447	0.451

注：括弧内は White の不均一分散一致標準誤差。*** は1%，** は5%，* は10% 水準で有意であることを示す。

パネル C　被説明変数「環境スコア」

説明変数	被説明変数　環境					
	(1)	(2)	(3)	(4)	(5)	(6)
探索財	−4.497 (6.42)					−8.550 (6.20)
経験財 [耐久性なし]		5.404** (2.56)				1.485 (2.53)
経験財 [耐久性あり]			3.274 (3.29)			1.208 (3.25)
経験財 [サービス]				−17.968*** (3.07)		−18.119*** (3.20)
信頼財					−11.659*** (2.14)	−14.136*** (2.85)
ln 総資産	8.373*** (0.33)	8.384*** (0.33)	8.286*** (0.35)	8.533*** (0.32)	8.382*** (0.33)	8.451*** (0.34)
R&D 集約度	70.359*** (25.26)	74.792*** (25.63)	61.814** (27.27)	40.281* (22.59)	71.546*** (25.27)	34.518 (24.90)
売上高 広告費比率	−69.567* (37.27)	−91.302** (38.02)	−83.477** (36.16)	−17.212 (33.18)	−74.449** (36.32)	−14.729 (37.68)
_cons	−39.845*** (3.96)	−40.354*** (3.97)	−38.876*** (4.22)	−40.035*** (3.91)	−39.999*** (3.98)	−38.971*** (4.14)
観測数	621	621	621	621	621	621
adjusted R2	0.481	0.483	0.481	0.520	0.480	0.523

パネル D　被説明変数「企業統治・社会性スコア」

説明変数	被説明変数　企業統治・社会性					
	(1)	(2)	(3)	(4)	(5)	(6)
探索財	0.598 (2.84)					1.326 (3.04)
経験財 [耐久性なし]		6.444*** (1.66)				6.397*** (1.79)
経験財 [耐久性あり]			1.991 (2.31)			2.556 (2.36)
経験財 [サービス]				−4.211* (2.28)		−2.757 (2.33)
信頼財					25.803*** (4.11)	25.626*** (4.22)
ln 総資産	7.445*** (0.30)	7.435*** (0.29)	7.379*** (0.31)	7.476*** (0.29)	7.460*** (0.29)	7.399*** (0.30)
R&D 集約度	39.742* (22.54)	43.262* (22.80)	33.557 (23.50)	32.204 (22.28)	39.887* (22.41)	31.416 (24.33)
売上高 広告費比率	68.834*** (24.59)	49.628** (24.73)	64.126** (25.23)	82.963*** (24.38)	69.105*** (24.07)	49.785* (28.34)
_cons	−28.087*** (3.49)	−28.330*** (3.46)	−27.294*** (3.63)	−28.031*** (3.46)	−28.318*** (3.46)	−27.690*** (3.59)
観測数	621	621	621	621	621	621
adjusted R2	0.521	0.527	0.521	0.524	0.523	0.531

注：括弧内は White の不均一分散一致標準誤差。*** は1%，** は5%，* は10% 水準で有意であることを示す。

［耐久性あり］では，経験財［耐久性あり］の方がプラス方向に係数の値が大きいという傾向は確認できなかった。

6-1-2. 消費財供給企業だけをサンプルとするOLS推定（N=228）

次に，先ほどのサンプル企業（621社）から中間財だけを供給する企業（393社）を除き，同様の回帰分析を行う。これにより，サンプル企業は228社となる。ここでのサンプル企業は，5種類の財・サービスのタイプに分類可能なセグメントを一つ以上持つ企業となる。

分析の結果は，表8のパネルAからDにまとめた通りである。パネルAはCSRスコアを，パネルBからDは雇用スコア，環境スコア，企業統治＋社会性スコアをそれぞれ被説明変数として分析を行った結果である。各パネルの(1)から(5)式は，5つある財・サービスのタイプのうち，単独の財・サービスを説明変数とし，これにコントロール変数を加えてOLS推定を行った。そして，(6)式は，すべての財・サービスのタイプを説明変数とし，すべてのコントロール変数とともにOLS推定を行った。

観察された事実は次の通りである。購入後に財の価値を評価できる経験財（経験財［耐久性なし］，経験財［耐久性あり］）の係数は，おおむねプラスで有意であり，経験財はCSR活動と有意に正の関係が検出された。さらに，被説明変数を雇用スコアとした場合を除き，購入頻度が少ない経験財［耐久性あり］と購入頻度が多い経験財［耐久性なし］では，経験財［耐久性あり］の方が係数の値が大きい傾向も確認できる。

表8パネルA，(6)式では経験財［耐久性なし］と経験財［耐久性あり］の係数の値は，それぞれ19.389と24.094となっている。この係数の値は，それぞれの財のタイプに該当する消費財の供給をゼロから50％に増加させたとき，CSRスコアを，経験財［耐久性なし］では9.7点，そして，経験財［耐久性あり］では12.0点上昇させるインパクトを持つことを意味する。

他方で，表7での推計結果と同様に，経験財［サービス］の係数はおおむねマイナスで有意であり，信頼財の係数は，被説明変数を環境スコアとした場合を除き，プラスで有意という傾向も確認された。また，企業規模，売上高広告費比率においては，おおむねプラスで有意な係数の値がみられた。

表8　消費財供給企業だけをサンプルとする OLS 推定 (N=228)

パネル A　被説明変数「CSR スコア」

説明変数	(1)	(2)	(3)	(4)	(5)	(6)
探索財	−4.934 (9.46)					−1.144 (10.98)
経験財 [耐久性なし]		17.785*** (5.13)				19.389*** (7.09)
経験財 [耐久性あり]			21.174** (8.52)			24.094** (9.89)
経験財 [サービス]				−25.265*** (6.00)		−12.615 (7.68)
信頼財					50.429*** (7.28)	46.961*** (46.96)
ln 総資産	23.192*** (1.27)	23.683*** (1.26)	22.217*** (1.32)	22.949*** (1.25)	23.423*** (1.25)	22.481*** (1.30)
R&D 集約度	32.686 (71.58)	47.651 (72.12)	−60.325 (77.11)	−30.726 (65.93)	34.819 (71.40)	−91.818 (73.85)
売上高 広告費比率	172.256** (73.10)	148.697** (68.69)	155.778** (72.26)	184.224*** (68.69)	172.825** (72.01)	143.568** (70.15)
_cons	−104.406*** (15.23)	−113.978*** (15.29)	−95.124*** (15.59)	−94.314*** (15.55)	−107.833*** (14.94)	−98.929*** (15.90)
観測数	228	228	228	228	228	228
adjusted R2	0.598	0.611	0.611	0.628	0.600	0.645

パネル B　被説明変数「雇用スコア」

説明変数	(1)	(2)	(3)	(4)	(5)	(6)
探索財	0.461 (3.57)					3.830 (4.44)
経験財 [耐久性なし]		4.463* (2.47)				6.998** (3.52)
経験財 [耐久性あり]			2.905 (3.32)			5.900 (4.13)
経験財 [サービス]				−3.739 (3.00)		0.877 (3.86)
信頼財					32.924*** (3.65)	35.440*** (4.62)
ln 総資産	6.973*** (0.58)	7.069*** (0.58)	6.821*** (0.61)	6.918*** (0.58)	7.064*** (0.56)	7.013*** (0.59)
R&D 集約度	1.836 (29.08)	4.959 (29.08)	−11.348 (32.41)	−7.990 (28.67)	1.859 (28.81)	−15.696 (32.64)
売上高 広告費比率	95.785*** (28.43)	90.974*** (28.05)	94.261*** (28.53)	98.327*** (29.20)	98.542*** (27.98)	84.009*** (29.74)
_cons	−30.911*** (7.26)	−32.896*** (7.38)	−29.358*** (7.40)	−29.125*** (7.35)	−32.244*** (6.98)	−34.053*** (7.67)
観測数	228	228	228	228	228	228
adjusted R2	0.395	0.402	0.397	0.400	0.403	0.417

注：括弧内は White の不均一分散一致標準誤差。*** は1%，** は5%，* は10% 水準で有意であることを示す。

パネル C　被説明変数「環境スコア」

説明変数	(1)	(2)	(3)	(4)	(5)	(6)
探索財	-2.805 (6.20)					-5.422 (6.80)
経験財 [耐久性なし]		9.265*** (2.72)				6.205* (3.60)
経験財 [耐久性あり]			15.130*** (4.27)			12.455** (4.88)
経験財 [サービス]				-18.678*** (3.28)		-14.438*** (4.04)
信頼財					-4.814 (3.33)	-12.547** (4.94)
ln 総資産	8.703*** (0.58)	8.963*** (0.58)	7.995*** (0.61)	8.510*** (0.58)	8.734*** (0.59)	7.962*** (0.60)
R&D 集約度	39.485 (25.95)	47.368* (26.31)	-27.245 (27.39)	-7.707 (20.17)	40.500 (25.93)	-50.057** (24.47)
売上高 広告費比率	1.474 (43.99)	-10.951 (41.97)	-9.834 (41.20)	10.868 (36.11)	-0.703 (43.43)	-4.018 (36.62)
_cons	-47.373*** (7.34)	-52.417*** (7.46)	-40.563*** (7.58)	-39.705*** (7.70)	-47.850*** (7.48)	-36.464*** (8.09)
観測数	228	228	228	228	228	228
adjusted R2	0.441	0.460	0.475	0.523	0.441	0.547

パネル D　被説明変数「企業統治・社会性スコア」

説明変数	(1)	(2)	(3)	(4)	(5)	(6)
探索財	-0.987 (2.99)					1.398 (3.78)
経験財 [耐久性なし]		6.070*** (1.81)				7.550*** (2.73)
経験財 [耐久性あり]			5.849* (2.99)			7.920** (3.65)
経験財 [サービス]				-7.208*** (2.45)		-2.331 (3.09)
信頼財					23.859*** (3.95)	24.308*** (5.31)
ln 総資産	7.162*** (0.49)	7.318*** (0.49)	6.887*** (0.53)	7.086*** (0.49)	7.249*** (0.48)	7.024*** (0.51)
R&D 集約度	10.724 (26.51)	15.572 (26.83)	-15.106 (28.37)	-7.524 (24.74)	11.232 (26.33)	-24.168 (27.95)
売上高 広告費比率	69.325** (28.37)	61.737** (27.37)	65.017** (28.67)	73.012*** (27.90)	70.467** (28.01)	57.383** (28.38)
_cons	-23.428*** (5.92)	-26.524*** (5.86)	-20.772*** (6.14)	-20.446*** (6.17)	-24.719*** (5.70)	-23.663*** (6.42)
観測数	228	228	228	228	228	228
adjusted R2	0.476	0.490	0.485	0.497	0.480	0.516

注：括弧内は White の不均一分散一致標準誤差。*** は1%，** は5%，* は10% 水準で有意であることを示す。

4 消費財の情報特性が CSR 活動に与える影響の分析　317

以上に示した (1) 経験財の係数がプラスで有意となっていたという事実は，仮説1と整合的であり，(2) 経験財［耐久性なし］と経験財［耐久性あり］では，経験財［耐久性あり］の方がおおむねプラス方向に係数の値が大きい傾向を持つという事実は，仮説2と整合的なものである。

7. 頑健性の検証

本節では，サンプルセレクションバイアスに対処した推計と，分類Ⅱの説明変数を用いた推計を行い，前節で述べた推計結果の頑健性の検証を行う。

7-1. サンプルセレクションバイアスへの対処

基本推計には，後述する2つの要因により，サンプルセレクションバイアスが生じている懸念がある。したがって次に，サンプルセレクションバイアスを考慮した推計を行い，頑健性を検証する必要がある。

東洋経済新報社「第6回CSR企業ランキング」は，先に述べたように，CSR評価（300点満点）と財務評価（300点満点）の2項目が同じウェイトで評価され，ランキングに入るかどうかが決まる。このため，CSR評価ではA社＜B社という2社があるとき，A社はCSR評価30点，財務評価60点の計90点でランキング入りし，B社はCSR評価40点，財務評価45点の計85点でランキング入りしない，というCSR評価についての逆転現象が生じうる。

また，CSRスコアが観察可能なのはランキングに含まれた730社に限定されることも問題を引き起こす。CSRスコアが観察可能なサンプル群を対象とした推計が，全上場企業をサンプル群とした推計と同様の傾向を示すとは限らず，CSRスコアが観察不可能な企業群を考慮に入れた推計を行う必要が生じる。

このようなサンプルセレクションバイアスへの対処として，ヘックマンの二段階推定（Heckman, 1976；Heckman, 1979）を用いる。縄田（1993），Nawata（1994），縄田（1997），Siegel and Vitaliano（2007）を参考として以下の推計モデルを定式化する。

$$\text{RANK-IN}_{it} = a_0 + a_1 \text{ROA}_{it-1} + a_2 \text{AGE}_{it-1} + a_3 \text{DASS}_{it-1} + a_4 \text{FRGN}_{it-1} + \varepsilon^* \quad (2)$$
$$\text{CSR}_{it} = a_0 + a_1 \text{GOODTYPE}_{it-1} + a_2 \text{ASS}_{it-1} + a_3 \text{RD}_{it-1} + a_4 \text{ADV}_{it-1} + a_5 \lambda + \varepsilon \quad (3)$$

添字iは企業iを,添字tは時点tを示す。ε^* と ε は,それぞれの推計式における誤差項を表す。

先行研究ではヘックマンの二段階推定を行う際,一段階目に導入される変数が二段階目で用いられる変数と同じ（あるいは,その一部を含む）モデルでは,計量上の重大な問題が生じる恐れがあることを指摘している（縄田, 1993；Nawata, 1994；縄田, 1997）。この指摘を踏まえ,本研究では一段階目と二段階目とでは異なる変数を導入した。

一段階目（(2) 式：選択モデル）の被説明変数はCSRランキング入りした企業を1,それ以外の企業を0とする離散変数であり,これを収益性,成熟度,財務安全性,外部圧力という変数により回帰する（プロビット推定）。この選択モデルを推計することにより,逆ミルズ比（標準正規分布の確率密度関数／標準正規分布の累積分布関数）が計算される。

二段階目（(3) 式）では,基本推計と同様の推計式に,逆ミルズ比を加えた分析を行う。逆ミルズ比は,CSRランキング入りする確率とほぼ同義と考えることができる。そのため,この二段階目において逆ミルズ比を導入することで,サンプルセレクションバイアスへの対処がなされた推計結果を得ることが可能となる。

ROAは,Siegel and Vitaliano (2007) に従い,収益性の代理変数として,一期前の業種等調整済みのROAを導入した。AGEは,Suto and Takehara (2013) を参考に,成熟度の代理変数として,2010年までの企業の創業年数を導入した。DASSは,Bae et al. (2011),および,佐々木・花枝 (2014) を参考に,財務安全度の代理変数として,一期前の負債比率を導入した。そして,FRGNは,Tanimoto and Suzuki (2005),および,Suzuki et al. (2010) を参考として,一期前の外国人持株比率を導入した[10]。

4 消費財の情報特性が CSR 活動に与える影響の分析　319

7-2. 説明変数を分類Ⅱとする分析

続いて，その他の変数は基本推計と同一とし，説明変数だけを分類Ⅰから分類Ⅱの財・サービスの分類とした推計モデルにより，頑健性の検証を行った。

分析の結果は，表9のパネルBに示した通りである。(1), (2) 式はCSRスコアを，(3), (4) 式は雇用スコアを，(5), (6) 式は環境スコアを，(7), (8) 式は企業統治＋社会性スコアをそれぞれ被説明変数にして，すべての財・サービスのタイプを導入し，コントロール変数のもとOLS推定を行った。また，(1), (3), (5), (7) 式はサンプル企業をCSRデータが入手可能な621社とし，(2), (4), (6), (8) 式は上記のサンプル企業（621社）から，中間財だけを供給する企業（383社）を除き[11]，同様の回帰分析を行った。

7-3. 検証の結果

表9のパネルAでは，サンプルセレクションバイアスへの対処を行った。逆ミルズ比に注目すると，雇用を被説明変数とする (2) 式だけが，統計的に有意となっていることが分かる。したがって，実際にサンプルセレクションバイアスが生じているのは，雇用スコアを被説明変数とするケースのみであり，それ以外ではサンプルセレクションバイアスが生じていない。サンプルセレクションバイアスを考慮した推計から明らかとなったのは，以下の事実である。

環境スコアを被説明変数とするケースを除き，経験財［耐久性なし］の係数はプラスで有意であった。また，環境スコアを被説明変数とする推計式において，探索財の係数はマイナスで有意となっていた。以上の傾向は，基本推計（サンプル企業を621社とした推計）と，ほぼ同様の結果となった。

表9のパネルBでは，説明変数を分類Ⅱに変更して頑健性の検証を行った。サンプル企業を621社とする (1), (3), (5), (7) 式からは経験財［耐久性なし］の係数の値がプラスで有意であるという結果が示された。これは，基本推計（サンプル企業を621社とした推計），ならびに，ヘックマンの二段階推定から示された傾向とほぼ合致する結果である。また，サンプル企業を238社とする (2), (4), (6), (8) 式からは経験財［耐久性なし］と経験財［耐久性あり］の係数の値がおおむねプラスで有意であるという傾向が確認された。この

表9 頑健性の検証
パネルA ヘックマンの二段階推定（サンプルセレクションバイアスへの対処）

説明変数	被説明変数			
	CSR (1)	雇用 (2)	環境 (3)	企業統治・社会性 (4)
探索財	−6.839 (10.49)	2.716 (4.80)	−8.831* (5.16)	1.079 (4.35)
経験財 [耐久性なし]	11.683** (5.70)	5.478** (2.61)	1.389 (2.80)	6.312*** (2.36)
経験財 [耐久性あり]	4.848 (5.98)	0.932 (2.75)	1.124 (2.94)	2.483 (2.48)
経験財 [サービス]	−16.471*** (5.32)	1.450 (2.44)	−17.501*** (2.62)	−2.211 (2.21)
信頼財	42.760 (40.10)	35.960** (18.26)	−13.651 (19.72)	26.054 (16.61)
ln 総資産	21.319*** (0.98)	6.146*** (0.45)	8.015*** (0.48)	7.014*** (0.40)
R&D 集約度	79.980* (42.84)	29.451 (19.69)	29.482 (21.07)	26.972 (17.76)
売上高広告費比率	101.847 (71.28)	68.738** (32.68)	−16.214 (35.06)	48.474 (29.55)
_cons	−72.983*** (15.32)	−16.519** (7.05)	−30.365*** (7.53)	−20.097*** (6.35)
[選択モデル]				
ROA	0.011** (0.004)	0.011** (0.004)	0.011** (0.004)	0.011** (0.004)
創業年数	0.013*** (0.001)	0.013*** (0.001)	0.013*** (0.001)	0.013*** (0.001)
負債比率	0.001 (0.001)	0.001 (0.001)	0.001 (0.001)	0.001 (0.001)
外国人持株比率	0.036*** (0.002)	0.036*** (0.002)	0.036*** (0.002)	0.036*** (0.002)
_cons	−1.941*** (0.10)	−1.941*** (0.10)	−1.941*** (0.10)	−1.941*** (0.10)
逆ミルズ比 λ	−5.645 (4.12)	−5.334*** (1.91)	−2.829 (2.03)	−2.496 (1.71)
観測数（①+②）	3013	3013	3013	3013
①打ち切りデータ	2392	2392	2392	2392
②非打ち切りデータ	621	621	621	621
Wald chi2 (8)	520.00	226.47	326.22	343.06
Prob>chi2	0.000	0.000	0.000	0.000

注：括弧内は標準誤差。*** は1%，** は5%，* は10% 水準で有意であることを示す。

パネルB　OLS推定（説明変数：分類Ⅱ）

説明変数	被説明変数 CSR (1)	(2)	被説明変数 雇用 (3)	(4)	被説明変数 環境 (5)	(6)	被説明変数 企業統治・社会性 (7)	(8)
探索財	−12.683 (8.26)	−5.982 (9.91)	2.105 (3.62)	1.604 (4.04)	−13.355*** (4.75)	−8.681 (6.08)	1.145 (3.10)	3.193 (3.55)
経験財 [耐久性なし]	14.692*** (4.09)	19.341*** (5.34)	6.129*** (1.74)	5.765** (2.52)	4.354** (2.19)	7.728*** (3.25)	6.022*** (1.50)	7.818*** (2.04)
経験財 [耐久性あり]	3.034 (3.98)	17.815*** (6.75)	1.472 (1.69)	4.046 (2.84)	1.555 (1.84)	10.150** (4.02)	0.543 (1.58)	6.047** (2.47)
経験財 [サービス]	−9.841** (3.80)	−2.584 (6.55)	−0.310 (1.91)	−0.547 (2.79)	−9.111*** (2.16)	−4.022 (4.07)	−1.676 (1.63)	1.120 (2.38)
信頼財	−4.652 (15.26)	−2.210 (15.82)	−0.030 (4.25)	1.669 (4.72)	−6.445 (8.09)	−5.091 (8.38)	3.599 (4.99)	6.543 (5.33)
ln 総資産	22.115*** (0.68)	21.327*** (1.14)	6.931*** (0.34)	6.221*** (0.61)	8.351*** (0.34)	7.983*** (0.62)	7.409*** (0.30)	6.422*** (0.50)
R&D 集約度	103.126 (67.34)	−50.105 (70.43)	36.170 (26.33)	−10.439 (31.16)	42.849* (24.55)	−29.359 (21.27)	38.842 (24.38)	−0.704 (28.77)
売上高広告費比率	92.936 (64.53)	91.091 (62.08)	71.598** (29.14)	70.770** (29.83)	−36.877 (36.00)	−37.582 (34.11)	54.520** (26.37)	49.144* (27.41)
_cons	−89.656*** (7.83)	−87.811*** (14.62)	−32.339*** (4.04)	−23.462*** (7.33)	−38.181*** (4.10)	−39.246*** (8.41)	−27.883*** (3.62)	−19.092*** (5.79)
観測数	621	238	621	238	621	238	621	238
adjusted R2	0.633	0.626	0.452	0.383	0.515	0.508	0.530	0.472

注：括弧内はWhiteの不均一分散一致標準誤差。***は1%，**は5%，*は10%水準で有意であることを示す。

結果は，基本推計（サンプル企業を228社とした推計）で観察された傾向と一致する。以上より，基本推計で計測された事実は，サンプルセレクションバイアスに対処した推計，あるいは，説明変数を分類Ⅱに変更した推計を行っても一定の頑健性を持つことが示された。

8. むすび

　本稿の目的は，CSR活動を行うという経営判断が，当該企業の供給する財・サービスのタイプによって決まることを（情報の非対称性の大きい産業ほどCSR活動に積極的であることを），日本企業を調査対象として検証することであった。日本企業を対象として，CSR活動への取り組み度合いを供給する財の特性との関連で分析した初めての試みである。

　消費財産業の商品特性が日本企業のCSR活動の実施に与える影響を分析するOLS推計からは，サンプル企業を621社とした場合には，経験財［耐久性なし］の係数は，おおむねプラスで有意であるという結果が観察された。経験財［耐久性なし］の係数がプラスであったという事実は，仮説1を部分的にサ

ポートするものであった。一方で，経験財［耐久性なし］と経験財［耐久性あり］では，経験財［耐久性あり］の方が係数の値が大きいという傾向は観察されなかった。

中間財だけを供給する企業（393社）をサンプル企業から除いた228社を対象とした同様のOLS推計からは，経験財（経験財［耐久性なし］，経験財［耐久性あり］）の係数はおおむねプラスで有意となっていて，経験財［耐久性なし］と経験財［耐久性あり］では，経験財［耐久性あり］の方が係数の値が大きいという傾向が観察された。経験財の係数がプラスであったという事実は，仮説1と整合的なものであり，経験財［耐久性なし］と経験財［耐久性あり］では，経験財［耐久性あり］の方が係数の値が大きいという傾向は，仮説2と整合的なものであった。

また，以上の事実は，CSRスコアの入手可能性についてのサンプルセレクションバイアスに対処した推計，あるいは，説明変数を分類IIに変更した推計といった，異なるアプローチのもとでの検証においても，その結果について一定の頑健性を持つことが示された。

しかしながら，本研究には次に述べるような留意点がある。本稿ではCSR活動の代理変数として東洋経済新報社「第6回CSR企業ランキング」を用いたが，Siegel and Vitaliano（2007）を含む米国企業を対象とする実証研究の多くは，KLD社の提供するCSRデータをCSR活動の代理変数として採用している。このCSRデータの違いが，推定結果に影響を与えた可能性は否定できない。日本企業を対象としたCSR活動の分析を行う際，この留意点を考慮することが課題となる。

(1) 首藤他（2006），首藤・竹原（2008a, 2008b），Suto and Takehara（2013）。
(2) Kitzmueller and Shimshack（2012）ではMcWilliams and Siegel（2001）とともに，理論モデルを用いた分析を行ったBaron（2001）を戦略的CSR理論の先駆的研究と整理している。
(3) 米国企業を対象とした分析でも同様の傾向を指摘する文献はある。例えば，Di Giuli and Kostovetsky（2014）ではKLDスコアと将来のROAとの間にマイナスの相関関係があることを示している。
(4) CSR活動を決定する要因には，産業特性（例えば，資本集約型産業と労働集約型産業）

等の要素も影響を与えることが予想されるが，本研究では消費財の情報特性に注目し分析を行う．

(5) ASS, RD, および, ADV の値の入手を一期前としたのは，企業経営者が CSR 活動の実施についての意思決定を行う直前の時点での企業の状態を反映させるためである．

(6) 東洋経済新報社による CSR 調査は，2005 年以降，毎年実施されている．2011 年 6 月に実施された調査は，全上場企業と主要未上場企業，3644 社へ調査票を送付するアンケート調査を実施し，そのうち回答の得られた 1117 社（上場企業 1062 社，主要未上場企業 55 社）について，回答結果などを基として CSR 活動の数値化を行っている．

(7) BtoB 取引は，事業体と消費者との間での情報の非対称性は非常に大きいことが考えられる．ただ，5 つの財・サービスのタイプへの分類は，消費者による財・サービスの質を知覚するタイミングが重要な判断要素となるため，BtoB 取引をこれらの分類に含むことは適当でないと判断した．この分類の基準については，今後，検討が必要である．

(8) セグメント情報は企業財務データバンクを利用し一期前のデータを使用した．当該年度のデータが入手できない場合は，前後のセグメント情報を入手可能な年度，または日経テレコン 21 から情報を入手している．

(9) 表 2 には「クリーニング」，「ゴミ収集」，「造園」など，Siegel and Vitaliano (2007), p. 780 での分類には含まれているが，日本の上場企業には該当のなかった業種は掲載していない．一方で，「結婚相談所」などの分類されていない業種や他のタイプの財・サービスとして分類するのが妥当と思われる業種は独自に分類を実施している．

(10) ROA, AGE, DASS, および, FRGN の値の入手を一期前としたのは，企業経営者が CSR 活動の実施についての意思決定を行う直前の時点での企業の状態を反映させるためである．

(11) ここではタイプⅡの財・サービスの分類を行っている．そのため，中間財だけを供給する企業の数が基本推計と異なる．タイプⅠでは 393 社が，タイプⅡでは 383 社が中間財だけを供給する企業であった．

＜**参考文献**＞

佐々木隆文・花枝英樹 (2014)「従業員処遇と資本構成」，『現代ファイナンス』，No. 35, 63-86 頁．

首藤惠 (2008)「CSR とコーポレート・ガバナンス」，宮島英昭編『企業統治分析のフロンティア』，日本評論社，212-236 頁．

―― (2012)「CSR 研究の新たなステージ：ビジネス・モデルと資本コスト」，『証券アナリストジャーナル』，第 50 巻第 9 号，54-60 頁．

――・竹原均 (2008a)「企業の社会的責任とコーポレート・ガバナンス (上)：非財務情報開示とステークホルダー・コミュニケーション」，『証券経済研究』，第 62 号，27-46 頁．

――・―― (2008b)「企業の社会的責任とコーポレート・ガバナンス (下)：非財務情報開示

とステークホルダー・コミュニケーション」,『証券経済研究』, 第63号, 29-49頁。
――・増子信・若園智明（2006）「企業の社会的責任（CSR）への取組みとパフォーマンス：企業収益とリスク」,『証券経済研究』, 第56号, 31-51頁。
東洋経済オンライン「CSR企業ランキング2012・トップ700」Available at http://toyokeizai.net/articles/-/8766/ Accessed 2014年12月16日。
縄田和満（1993）「タイプⅡのトービット・モデルの推定について」,『日本統計学会誌』, 第23巻第2号, 223-247頁。
――（1997）「Probit, Logit, Tobit」, 蓑谷千凰彦・廣松毅監修『応用計量経済学Ⅱ』, 多賀出版, 237-298頁。
広田真一（2012）『株主主権を超えて：ステークホルダー型企業の理論と実証』, 東洋経済新報社。
Aoki, M.（2010）*Corporations in Evolving Diversity: Cognition, Governance, and Institutions*, Oxford: Oxford University Press.（谷口和弘訳『コーポレーションの進化多様性：集合認知・ガバナンス・制度』, NTT出版, 2011年）
Bae, K. H., J. K. Kang and J. Wang（2011）"Employee Treatment and Firm Leverage: A Test of the Stakeholder Theory of Capital Structure," *Journal of Financial Economics*, Vol. 100, Issue 1, pp. 130-153.
Baron, D. P.（2001）"Private Politics, Corporate Social Responsibility, and Integrated Strategy," *Journal of Economics & Management Strategy*, Vol. 10, Issue 1, pp. 7-45.
Bénabou, R. and J. Tirole（2010）"Individual and Corporate Social Responsibility," *Economica*, Vol. 77, No. 305, pp.1-19.
Di Giuli, A. and L. Kostovetsky（2014）"Are Red or Blue Companies More Likely to Go Green? Politics and Corporate Social Responsibility," *Journal of Financial Economics*, Vol. 111, Issue 1, pp. 158-180.
Friedman, M.（1970）"The Social Responsibility of Business is to Increase Its Profits" *New York Times Magazine*, September 13, pp. 122-126.
Heckman, J. J.（1976）"The Common Structure of Statistical Models of Truncation, Sample Selection and Limited Dependent Variables and a Simple Estimator for Such Models," *Annals of Economic and Social Measurement*, Vol. 5, No. 4, pp. 475-492.
――（1979）"Sample Selection Bias as a Specification Error," *Econometrica*, Vol. 47, No. 1, pp. 153-161.
Kitzmueller, M. and J. Shimshack（2012）"Economic Perspectives on Corporate Social Responsibility," *Journal of Economic Literature*, Vol. 50, Issue 1, pp. 51-84.
Liebermann, Y. and A. Flint-Goor（1996）"Message Strategy by Product-Class Type: A Matching Model," *International Journal of Research in Marketing*, Vol. 13, Issue 3, pp. 237-249.

McWilliams, A. and D. S. Siegel (2000) "Corporate Social Responsibility and Financial Performance: Correlation or Misspecification?," *Strategic Management Journal*, Vol. 21, Issue 5, pp. 603-609.

——— and——— (2001) "Corporate Social Responsibility: A Theory of the Firm Perspective," *Academy of Management Review*, Vol. 26, Issue 1, pp. 117-127.

Nawata, K. (1994) "Estimation of Sample Selection Bias Models by the Maximum Likelihood Estimator and Heckman's Two-Step Estimator," *Economics Letters*, Vol. 45, Issue 1, pp. 33-40.

Nelson, P. (1970) "Information and Consumer Behavior," *Journal of Political Economy*, Vol. 78, No. 2, pp. 311-329.

——— (1974) "Advertising as Information," *Journal of Political Economy*, Vol. 82, No. 4, pp. 729-754.

Siegel, D. S. and D. F. Vitaliano (2007) "An Empirical Analysis of the Strategic Use of Corporate Social Responsibility," *Journal of Economics & Management Strategy*, Vol. 16, Issue 3, pp. 773-792.

Suto, M. and H. Takehara (2013) "The Impact of Corporate Social Performance on Financial Performance: Evidence from Japan," *Waseda University Institute of Finance Working Paper Series*, WIF-13-003.

Suzuki, K., K. Tanimoto and A. Kokko (2010) "Does Foreign Investment Matter? Effects of Foreign Investment on the Institutionalization of Corporate Social Responsibility by Japanese Firms," *Asian Business & Management*, Vol. 9, No. 3, pp. 379-400.

Tanimoto, K. and K. Suzuki (2005) "Corporate Social Responsibility in Japan: Analyzing the Participating Companies in Global Reporting Initiative," *The European Institute of Japanese Studies Working Papers Series*, No. 208.

付表1 探索財,経験財,信頼財の定義

1) 探索財…購入前に品質を評価することが可能な財。 Ex) 衣類,家具
2) 経験財…購入後に使用(消費)したとき,品質を評価することが可能な財。
   - a 耐久性の低いタイプ…頻繁に購入される経験財。 Ex) 食品,化粧品
      消費者は短期間で財の品質を理解する。
   - b 耐久性の高いタイプ…頻繁に購入することが難しい経験財。 Ex) 自動車,家電
      消費者は長い期間をかけ財の品質を理解する。
   - c サービス財タイプ …サービスとして提供される経験財。 Ex) 小売業,不動産業
3) 信頼財…購入後であっても,品質を評価することが難しい財。 Ex) 投資助言

注:Nelson (1970, 1974), Liebermann and Flint-Goor (1996), Siegel and Vitaliano (2007) を参考にして作成。

付表2　東洋経済新報社「第6回 CSR 企業ランキング」評価項目一覧

雇用	1. 女性社員比率　2. 離職者状況　3. 残業時間　4. 外国人管理職人数　5. 女性管理職比率　6. 女性部長職以上比率　7. 女性役員の有無　8. ダイバーシティ推進の基本理念　9. ダイバーシティ尊重の経営方針　10. 多様な人材登用部署　11. 障害者雇用率（実績）　12. 障害者雇用率の目標値　13. 有給休暇取得率　14. 産休期間　15. 産休取得者　16. 育児休業取得者　17. 男性の育児休業取得者　18. 配偶者の出産休暇制度　19. 介護休業取得者　20. 退職した社員の再雇用制度の有無　21. ユニークな両立支援制度　22. 勤務形態の柔軟化に関する諸制度　23. 従業員のインセンティブを高めるための諸制度　24. 労働安全衛生マネジメントシステム　25. 労働安全衛生分野の表彰歴　26. 労働災害度数率　27. 人権尊重等の方針　28. 人権尊重等の取り組み　29. 従業員の満足度調査　30. 新卒入社者の定着度
環境	1. 環境担当部署の有無　2. 環境担当役員の有無　3. 同役員の担当職域　4. 環境方針文書の有無　5. 同文書の第三者関与　6. 環境会計の有無　7. 同会計における費用と効果の把握状況　8. 同会計の公開状況　9. 環境監査　10. ISO 14001 取得体制　11. ISO 14001 取得率（国内）　12. ISO 14001 取得率（海外）　13. グリーン購入体制　14. 事務用品等のグリーン購入比率　15. グリーン調達体制　16. 環境ラベリング　17. 土壌・地下水の汚染状況把握　18. 環境関連法令違反の有無　19. 環境問題を引き起こす事故・汚染の有無　20. CO^2 排出量等削減への中期計画の有無　21. 気候変動への対応の取り組み　22. 環境対策関連の表彰歴　23. 環境ビジネスへの取り組み　24. 生物多様性保全への取り組み　25. 生物多様性保全プロジェクトへの支出額
企業統治	1. CSR 活動のマテリアリティ設定　2. ステークホルダー・エンゲージメント　3. CSR 担当部署の有無　4. CSR 担当役員の有無　5. 同役員の担当職域　6. CSR 方針の文書化の有無　7. IR 担当部署　8. 法令順守関連部署　9. 国内外の CSR 行動基準への参加等　10. 内部告発窓口設置　11. 内部告発者の権利保護に関する規定制定　12. 公正取引委員会など関係官庁からの排除勧告　13. 不祥事などによる操業・営業停止　14. コンプライアンスに関わる事件・事故での刑事告発　15. 汚職・贈収賄防止の方針　16. 政治献金等の開示　17. 内部統制の基本的な取り組み　18. 内部統制の評価　19. 情報システムに関するセキュリティポリシーの有無　20. 情報システムのセキュリティに関する内部監査の状況　21. 情報システムのセキュリティに関する外部監査の状況　22. プライバシー・ポリシーの有無　23. リスクマネジメント・クライシスマネジメントの状況　24. 企業倫理方針の文書化・公開　25. 倫理行動規定・規範・マニュアルの有無
社会性	1. 消費者対応部署の有無　2. 社会貢献担当部署の有無　3. 商品・サービスの安全性・安全体制に関する部署の有無　4. 社会貢献活動支出額　5. NPO・NGO 等との連携　6. ESG 情報の開示　7. SRI インデックス等への組み入れ・エコファンド等の採用状況　8. 消費者からのクレーム等への対応マニュアルの有無　9. 同クレームのデータベースの有無　10. ISO 9000S の取得状況（国内）　11. ISO 9000S の取得状況（海外）　12. ISO 9000S 以外の品質管理体制　13. 地域社会参加活動実績　14. 教育・学術支援活動実績　15. 文化・芸術・スポーツ活動実績　16. 国際交流活動実績　17. CSR 調達への取り組み状況　18. ボランティア休暇　19. ボランティア休職　20. マッチング・ギフト　21. BOP ビジネスの取り組み　22. 海外での CSR 活動　23. CSR 関連の表彰歴

出所：東洋経済オンライン「CSR 企業ランキング2012・トップ700」より作成。

【企業と社会シリーズ4】

持続可能性と戦略　Sustainability and Strategy

2015年9月1日　発行

編　者　企業と社会フォーラム
発行者　千倉成示
発行所　株式会社　千倉書房
　　　　〒104-0031　東京都中央区京橋2-4-12
　　　　Tel 03-3273-3931　Fax 03-3273-7668
　　　　http://www.chikura.co.jp/

印刷／製本　藤原印刷

|JCOPY|〈(社)出版者著作権管理機構　委託出版物〉

本書のコピー，スキャン，デジタル化など無断複写は著作権法上での例外を除き禁じられています。複写される場合は，そのつど事前に，㈳出版者著作権管理機構(電話03-3513-6969，FAX03-3513-6979，e-mail: info@jcopy.or.jp)の許諾を得てください。また，本書を代行業者などの第三者に依頼してスキャンやデジタル化することは，たとえ個人や家庭内での利用であっても一切認められておりません。

ISBN978-4-8051-1070-6